Wetland Development in Paddy Fields and Disaster Management

AF173012

Akira Matsui

Wetland Development in Paddy Fields and Disaster Management

 Springer

Akira Matsui
Matsui Store Co., Ltd.
Obama, Fukui, Japan

ISBN 978-981-19-3737-8   ISBN 978-981-19-3735-4   (eBook)
https://doi.org/10.1007/978-981-19-3735-4

© The Editor(s) (if applicable) and The Author(s), under exclusive license to Springer Nature
Singapore Pte Ltd. 2022
This work is subject to copyright. All rights are solely and exclusively licensed by the Publisher, whether
the whole or part of the material is concerned, specifically the rights of translation, reprinting, reuse
of illustrations, recitation, broadcasting, reproduction on microfilms or in any other physical way, and
transmission or information storage and retrieval, electronic adaptation, computer software, or by similar
or dissimilar methodology now known or hereafter developed.
The use of general descriptive names, registered names, trademarks, service marks, etc. in this publication
does not imply, even in the absence of a specific statement, that such names are exempt from the relevant
protective laws and regulations and therefore free for general use.
The publisher, the authors, and the editors are safe to assume that the advice and information in this book
are believed to be true and accurate at the date of publication. Neither the publisher nor the authors or
the editors give a warranty, expressed or implied, with respect to the material contained herein or for any
errors or omissions that may have been made. The publisher remains neutral with regard to jurisdictional
claims in published maps and institutional affiliations.

This Springer imprint is published by the registered company Springer Nature Singapore Pte Ltd.
The registered company address is: 152 Beach Road, #21-01/04 Gateway East, Singapore 189721,
Singapore

Preface

Extreme weather conditions have increased the risk of flood damage in Japan. Increases in flood disasters associated with global warming are occurring not only in Japan but also around the world. Land use practices must be reformed to protect cities from flood damage. I propose using green infrastructure to balance biodiversity conservation and flood control. Paddy fields can be used as green infrastructure.

Paddy fields hold water during heavy rains, protecting cities, and they can be referred to as paddy field dams. However, abandoned cultivated land is increasing in paddy fields due to an aging population and depopulation. Therefore, the flood control effects of paddy field dams are no longer utilized.

I propose combining abandoned cultivated land in one area and creating wetlands. The wetlands will not only hold water during the flood season but also provide habitat for aquatic animals and a place for children to learn about the environment. In addition, fish farming in wetlands could provide food for many people. Recreational activities such as fishing could also be conducted.

This proposal will help government officials across the world, especially those involved in urban and rural planning. It is expected that this concept will be supported by not only engineers but also biologists.

This book's primary theme was developed based on the following three books *Effects of Dam on Downstream Aquatic Community in Japan* and *Aquatic Animal Ecology in a Consolidated Paddy Field, Japan*, published by LAMBERT Academic Publishing and *Dam Construction, Paddy Farmland Consolidation and Aquatic Community in Japan* published by Tokyo Tosho Shuppan.

March 2022

Dr. Akira Matsui
Director, Matsui Store Co., Ltd.
Obama, Fukui, Japan

Acknowledgments

I would like to thank Yosuke Nishida, editor of Earth Sciences and Geography at Springer Nature, for providing me with the opportunity to publish. Thank you also to Tetsuji Nakabo, professor emeritus at Kyoto University; Kazumi Hosoya, professor emeritus at Kindai University; Kazumi Tanida, professor emeritus at Osaka Prefecture University; Miyoshi Ida, bureau chief at Shiga Science Teaching Material Research Committee; Hiroshi Watanabe, bureau chief at Yamasaki Research Institute; and Masahiro Deguchi, bureau chief at Township Promotion Association of Oriental White Stork for permission to quote from copyright sources. I acknowledge Sean Tsukida, at The Mt. Aoba Research Institute, for revising my English manuscript. I wish to express my gratitude to American Journal Experts for editing the English in my manuscript. I wish to thank my wife Masami Matsui and my sons Satoshi Matsui and Hajime Matsui for encouraging me. I am indebted to my parents Kinji Matsui and Kazuyo Matsui.

Contents

List of Boxes

Part I
Flood Control

Chapter 1
Characteristics of Flood Disasters

Abstract When comparing Japanese and foreign rivers, Japanese rivers are steeper. Therefore, the damage caused by flooding in Japan is likely to be greater than that caused by flooding abroad. Due to the effects of urbanization and global warming, the possibility of flooding is extremely high. The number of short-duration rainfall events exceeding 50 mm/h has increased, and these rainfall events have become localized, centralized and severe recently. Because of the influence of climate change associated with global warming, it is predicted that the frequency of heavy rainfall will increase approximately 1.4-fold. It is unequivocal that human influence has resulted in the warming of the atmosphere, oceans and land. Consequently, there is concern that a large-scale flood-related disaster will occur.

Keyword Flood disaster · Global warming · Rainfall · River · Urbanization

1.1 Comparison of Rivers in Japan and the World

Japanese rivers are short in length and have a steep gradient from upstream to downstream. Therefore, they are characterized by rapid flows into the sea. When it rains, river water levels suddenly rise, and the flood peak is reached in a short time (Fig. 1.1).

In London, the Thames River flows through the lowest part of the city. On the other hand, in Tokyo, many rivers flow through the highest parts of the city. Therefore, the damage caused by flooding in Tokyo is likely to be greater than that in London (Fig. 1.2).

1.2 Impact of Urbanization

Before development, rainwater penetrated the ground, and surface water mainly flowed into rivers. After development, because the impervious area covered with concrete increased, a large amount of rainwater began to flow into rivers in a short time. As a result, floods have become more frequent (Fig. 1.3).

© The Author(s), under exclusive license to Springer Nature Singapore Pte Ltd. 2022
A. Matsui, *Wetland Development in Paddy Fields and Disaster Management*,
https://doi.org/10.1007/978-981-19-3735-4_1

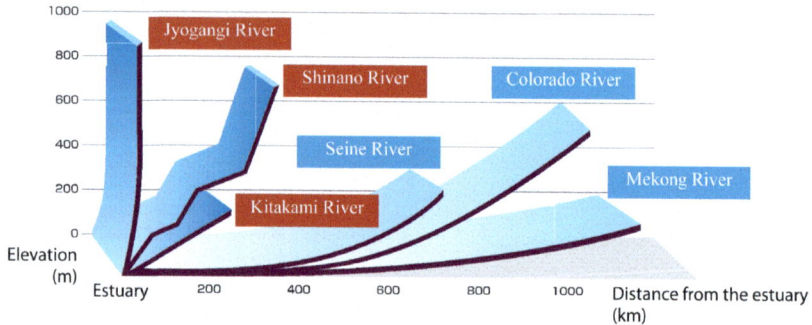

a : Jyoganji R., Shinano R. and Kitakami R.
b : Seine R., Colorado R. and Mekong R.

Fig. 1.1 Comparison of rivers in **a** Japan and **b** the world. *Source* Modified from the Ministry of Land, Infrastructure, Transport and Tourism of Japan, https://www.mlit.go.jp/river/pamphlet_jirei/kasen/gaiyou/panf/gaiyou2005/pdf/c1.pdf, Accessed October 30, 2021 (in Japanese)

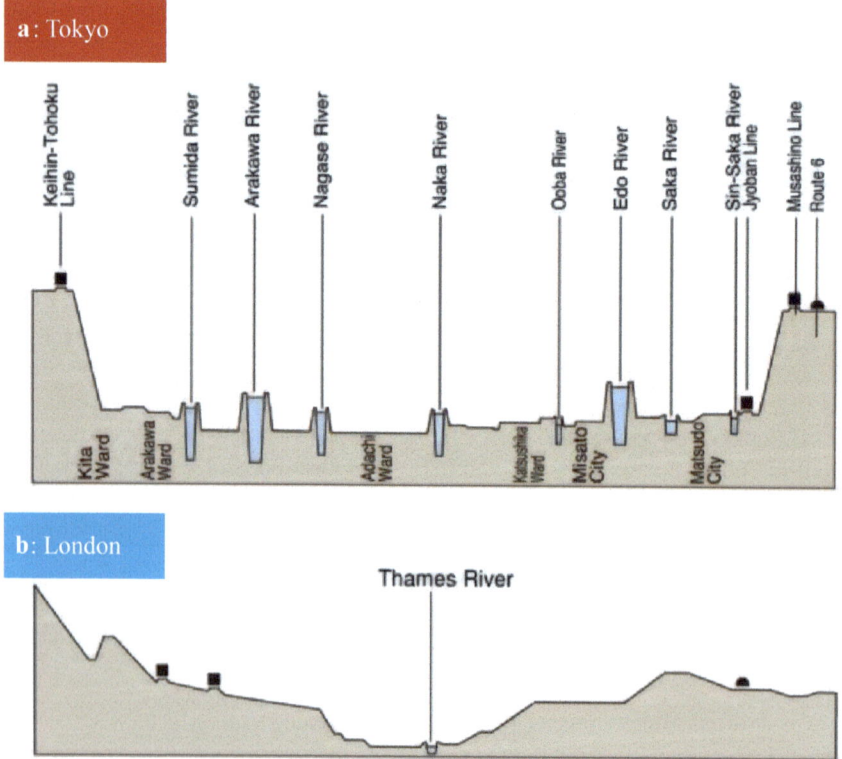

Fig. 1.2 Comparison of **a** the rivers in Tokyo and **b** the Thames R. in London. *Source* Modified from the Ministry of Land, Infrastructure, Transport and Tourism of Japan https://www.mlit.go.jp/river/basic_info/english/pdf/riversinjapan.pdf, Accessed May 18, 2022 (in Japanese)

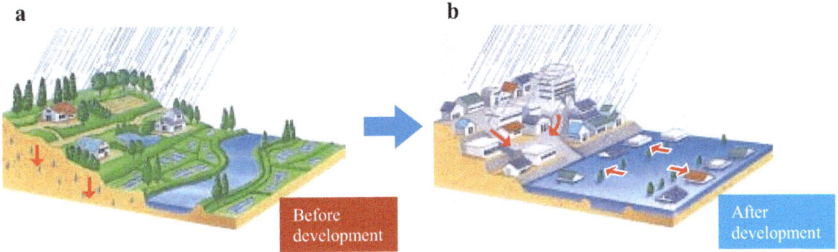

Fig. 1.3 Changes **a** before development and **b** after development. *Source* Modified from the Ministry of Land, Infrastructure, Transport and Tourism of Japan, https://www.mlit.go.jp/river/pam phlet_jirei/kasen/gaiyou/panf/gaiyou2005/pdf/c1.pdf, Accessed October 30, 2021 (in Japanese)

Global land use changes are four times greater than previously estimated (Karina et al. 2021), and such changes in land use likely affect climate change and cause disasters such as floods.

1.3 Global Warming and Flood-Related Disasters[1]

Human activities have influenced climate warming at a rate that is unprecedented in at least the last 2000 years (Fig. 1.4). It is unequivocal that human influence has resulted in the warming of the atmosphere, oceans and land. Widespread and rapid changes in the atmosphere, oceans, cryosphere and biosphere have occurred. Many changes in the climate system have increased in direct relation to increasing global warming. These changes include increases in the frequency and intensity of high temperature extremes, marine heatwaves, heavy precipitation, agricultural and ecological droughts in some regions, the proportion of intense tropical cyclones and reductions in Arctic sea ice, snow cover and permafrost (IPCC 2021).

Such temperature increases have also occurred in Japan (Fig. 1.5). The number of short-duration rainfall events exceeding 50 mm/h has increased, and these rainfall events have become localized, centralized and severe in recent years (Fig. 1.6). Because of the influence of climate change associated with global warming, it is predicted that the frequency of heavy rainfall event will increase approximately 1.4-fold. Consequently, there is concern that a large-scale flood-related disaster will occur. Rising temperatures are expected to have impacts on food and ecosystems and result in rising sea levels, increasing frequencies of heavy rainfall event, increasing tropical cyclone intensities and changing water availability in aquatic and coastal areas (Fig. 1.7).

---

[1] *Source* Reprinted from the UN Environment Programme, https://www.unep.org/news-and-stories/press-release/new-un-decade-ecosystem-restoration-offers-unparalleled-opportunity, Accessed October 30, 2021.

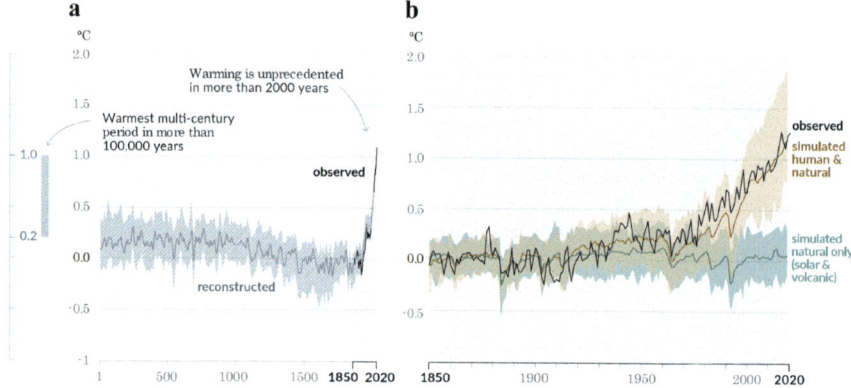

Fig. 1.4 History of change in global temperature and causes of recent warming. *Notes* **a** Change in global surface temperature (decadal average) as reconstructed (1–2000) and observed (1850–2020). **b** Change in global surface temperature (annual average) as observed and simulated using human and natural influences and only natural influences (both 1850–2020). *Source* Reprinted from IPCC (2021)

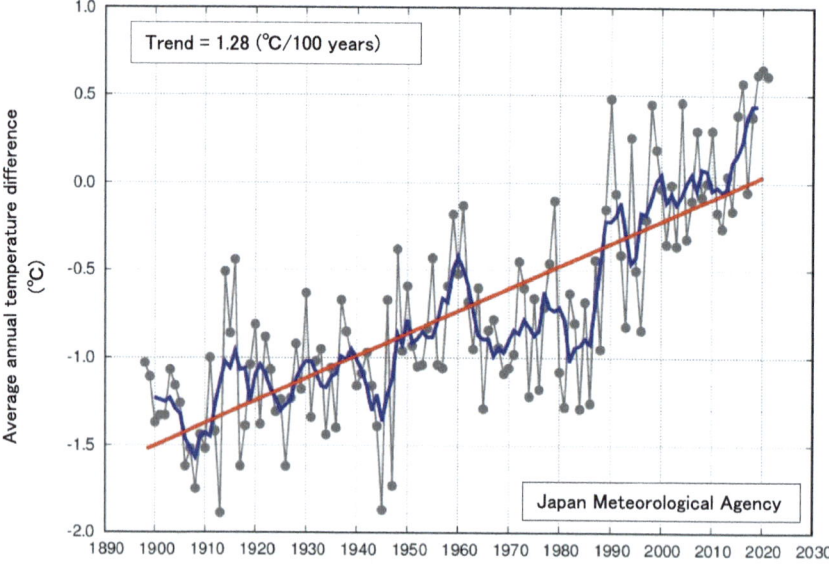

Fig. 1.5 Overall change in average annual temperature difference in Japan. *Note* The blue line shows the 5-year moving average, and the red line shows the long-term change trend. *Source* Modified from the Japan Meteorological Agency, https://www.data.jma.go.jp/cpdinfo/temp/an_jpn. html, Accessed May 18, 2022 (in Japanese)

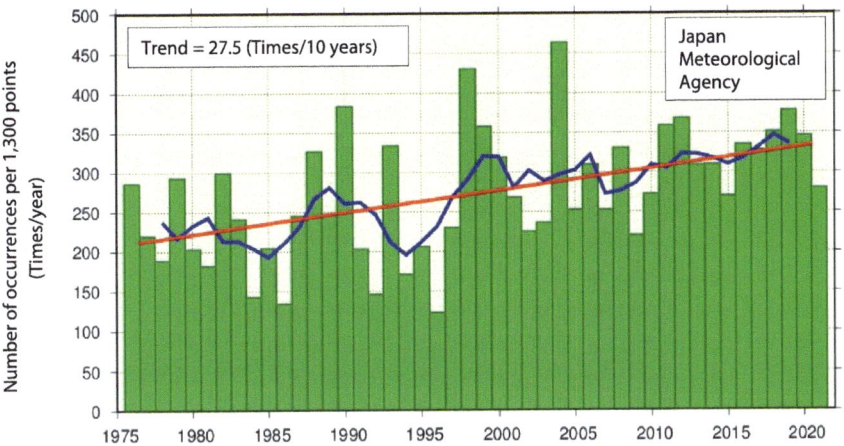

Fig. 1.6 Number of rainfall events exceeding 50 mm/h per year. *Note* The blue line shows the 5 year moving average, and the red line shows the long-term change trend. *Source* Modified from the Japan Meteorological Agency, https://www.data.jma.go.jp/cpdinfo/extreme/extreme_p.html, Accessed May 18, 2022 (in Japanese)

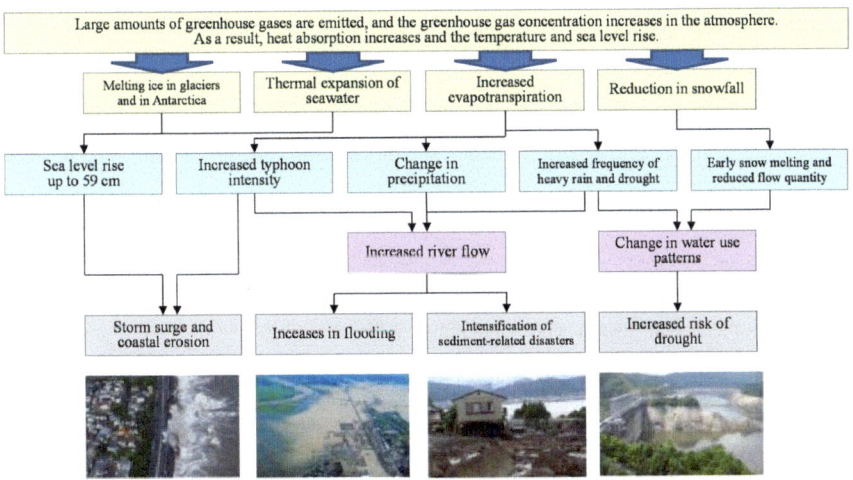

Fig. 1.7 Disasters caused by global warming. *Source* Modified from the Shinano River Ohkouzu Bosai Center (2009), http://www.hrr.mlit.go.jp/shinano/ohkouzu/bousaic/0912mizukanren/0912mizukanren.htm, Accessed October 30, 2021 (in Japanese)

Box 1.1 United Nations Designates 2021–2030 as Decade of Ecosystem Restoration

The General Assembly of the United Nations has announced that 2021 to 2030 will be designated as the decade of UN ecosystem restoration. Ecosystem restoration is the process of restoring degraded ecosystem functions, such as those of land, lakes and oceans, and increasing the productivity of ecosystems to the levels required by human society.

Deterioration of land and marine ecosystems has a negative impact on the well-being of 3.2 billion people, and the loss of biodiversity and ecosystem services accounts for approximately 10% of global gross domestic product. Approximately, 20% of the planted surface is eroded, depleted and polluted, reducing productivity by 2050, with global crop yields likely to decline by 10% and up to 50% in certain areas. Restoring 350 million hectares of degraded land by 2030 will generate $9 trillion in ecosystem services and remove 13–26 gigatons of greenhouse gases from the atmosphere.

This decade of ecosystem recovery will bring together political, scientific and financial support globally and lead to large-scale recovery from successful pilot efforts. The implementation of this effort will be led by the United Nations Environment Program (UNEP) and the Food and Agriculture Organization of the United Nations (FAO).

References

IPCC (2021) Climate Change 2021 the physical science basis summary for policymakers. In: Working group I contribution to the sixth assessment report of the intergovernmental panel on climate change

Winkler K, Fuchs R, Rounsevell M, Herold M (2021) Global land use changes are four times greater than previously estimated. Nat Commun 12:2501. https://doi.org/10.1038/s41467-021-22702-2

Chapter 2
Flood Control Methods

Abstract There are various flood control methods used in Japan. Because Japan has implemented modern hydraulic technology, to date, flood control methods have been used based on the concept that not even a drop of water will overflow from the outer bank to the inner bank. However, due to the effects of recent global warming, flood-related disasters have become more severe, and conventional flood control methods are no longer able to protect human lives and properties. Therefore, flood control methods must be changed to minimize damage from overflowing water from the outer bank to the inner bank. Thus, watershed control, open levees and paddy field dams should be implemented. An open levee is a traditional river construction method. Watershed control and paddy field dams are intended to be used as flood control methods in not only rivers but also the entire region.

Keywords Flood control · Modern hydraulic technology · Open levee · Paddy field dam · Traditional river construction method · Watershed control

2.1 Flood Control Methods

Flood control methods are shown in Fig. 2.1. These methods include the following: excavating a river channel to increase the cross section of water and lower the water level (Fig. 2.1a), building an embankment and enlarging the cross section of water (Fig. 2.1b), increasing the height of an embankment and enlarging the cross section of water (Fig. 2.1c), setting back the embankment to enlarge the cross section of water and lower the water level (Fig. 2.1d), constructing a dam to reduce the amount of water flowing to downstream rivers and lower the water level (Fig. 2.1e) and temporarily holding the floodwater at the detention basin when the water is about to overflow due to flooding (Fig. 2.1f).

While methods A to E are based on the idea that even a single drop of water should not overflow from the outer bank to the inner bank, method F assumes that water overflows from the outer bank to the inner bank. In comparison with the other methods, method F represents a substantial difference in flood control method, and method F is very important.

© The Author(s), under exclusive license to Springer Nature Singapore Pte Ltd. 2022
A. Matsui, *Wetland Development in Paddy Fields and Disaster Management*,
https://doi.org/10.1007/978-981-19-3735-4_2

a
River channel excavation

b
Embankment

c
Increasing the height of an embankment

d
Setting back an embankment

e
Dam

f
Detention basin

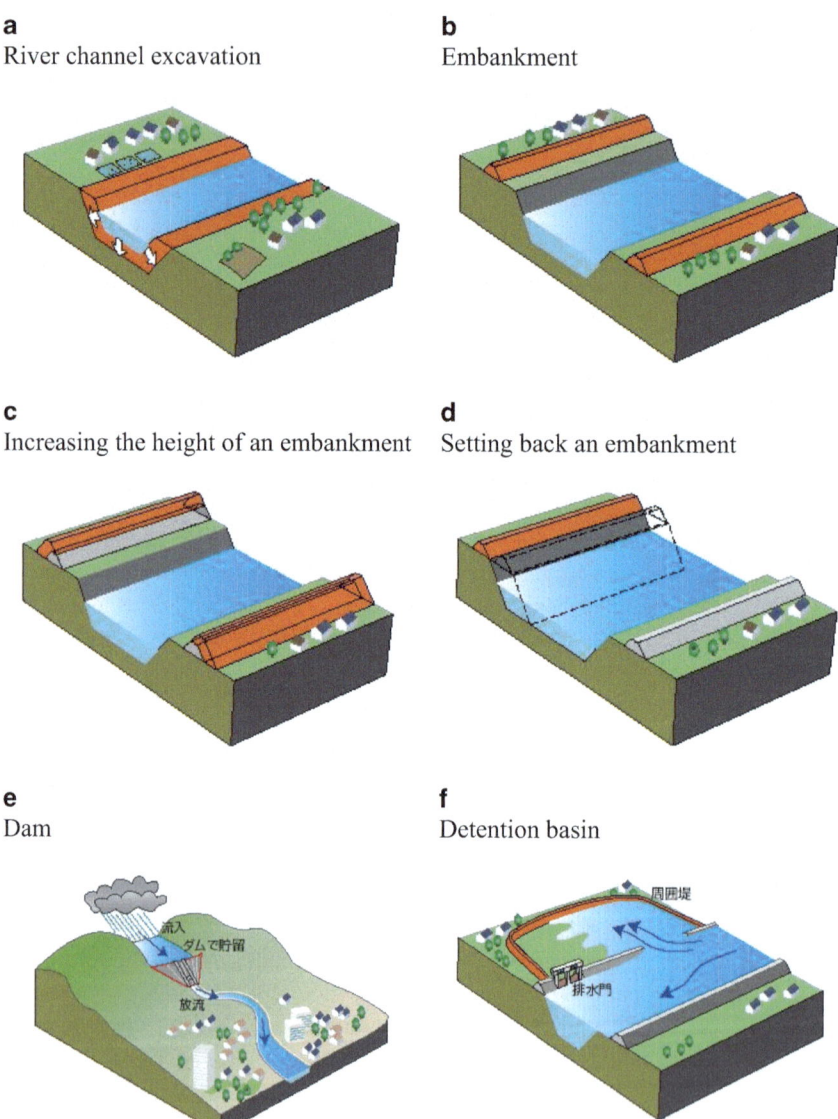

Fig. 2.1 Flood control methods. *Source* Modified from the Ministry of Land, Infrastructure, Transport and Tourism of Japan, https://www.mlit.go.jp/river/pamphlet_jirei/bousai/saigai/kiroku/suigai/suigai_4-5-ref2.html, Accessed October 30, 2021 (in Japanese)

a   **b**   **c**
Detention basin levee   Flood reduction levee   Drainage levee

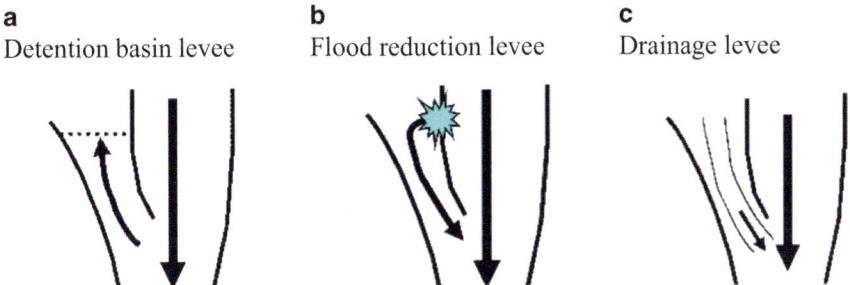

Fig. 2.2 Types of open levees. *Source* Modified from Teramura and Okuma (2005). Copyright 2005 Japan Society of Civil Engineers

2.2 Open Levee

An open levee is a form of river embankment with break in it (Fig. 2.2). As an open levee is discontinuous, water will overflow at the open part during a flood. However, the levees are constructed to overlap and contain the gaps; thus, flood-related disasters are limited to a certain area (Okuma 1987).

Open levees are classified into three types depending on their location and purpose. A detention effect can be obtained by temporarily causing the floodwaters at the opening to flow back into the levee (Fig. 2.3a). This levee type is used in gently sloping rivers. The flood reduction and drainage levee types are often used in steep rivers. The flood reduction levee type is designed on the assumption that there may be an upstream bank breakage (Fig. 2.3b). The drainage levee type is intended to remove inland water from tributaries (Fig. 2.3c).

An open levee is an autonomous structure created on the assumption that a flood will occur. However, flood control using a continuous embankment does not require a community to implement self-protection efforts, so local residents lose their ability to function autonomously (Teramura and Okuma 2005).

2.3 Paddy Field Dam

Runoff control devices are installed at the drainage outlets of paddy field plots to intentionally and temporarily store rainwater in paddy fields during intense rainfall events (Yoshikawa 2014). This measure is called a paddy field dam (Fig. 2.4). Paddy fields act as dams. Since there are many paddy fields in Japan, the dam effect of paddy fields is substantial.

a

b

Fig. 2.3 Open levee (*Photos* by Akira Matsui). *Notes* **a** Open levee 10 in Fig. 7.1. **b** Open levee 11 in Fig. 7.1. Photo date: February 26, 2021

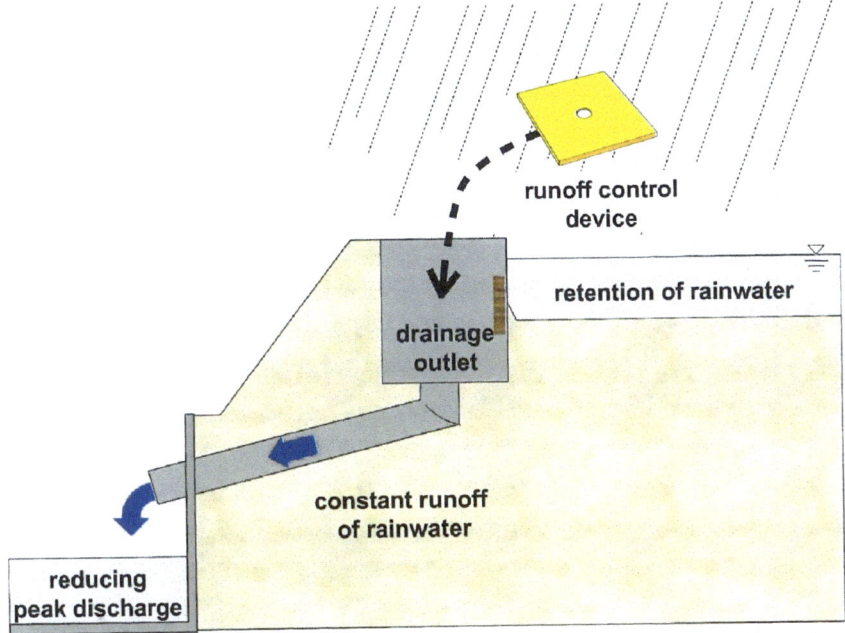

Fig. 2.4 Schematic of a paddy field dam. *Source* Reprinted with permission from Springer: Springer. Social-ecological restoration in paddy-dominated landscapes by Nishikawa Usio and Tadashi Miyashita 2014

2.4 Watershed Control

Watershed control involves embankment maintenance, dam construction/regeneration, etc. The concept of watershed control is that everyone in the catchment area of the basin where inundation is expected due to river flooding will take measures against flood-related disasters (Fig. 2.5).

As part of watershed control, sediment accumulated in rivers is removed to increase flow capacity (Figs. 2.6 and 2.7). In addition, embankments and slopes are strengthened (Fig. 2.8). To determine the water level that corresponds to flood risk, a water-level gauge is installed, and the river water level is observed (Fig. 2.9).

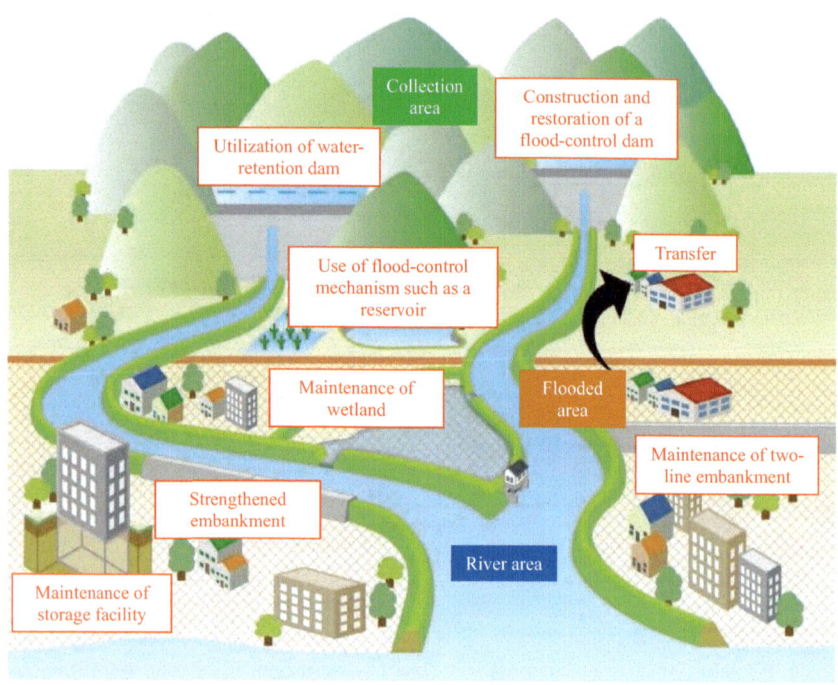

Fig. 2.5 Watershed control. *Source* Modified from the Ministry of Land, Infrastructure, Transport and Tourism of Japan, https://www.mlit.go.jp/report/press/content/001370162.pdf, Accessed October 30, 2021 (in Japanese)

Fig. 2.6 Removal of sediment accumulated in rivers—Case 1 (*Photos* by Akira Matsui). *Notes* The photos were taken at Fuchu Bridge on the Kita River. **a** Before sediment removal. Photo date: March 19, 2010. **b** After sediment removal. Photo date: March 11, 2021

a

Sediment

b

Fig. 2.7 Removal of sediment accumulated in rivers—Case 2 (*Photos* by Akira Matsui). *Notes* The photos were taken at Maruyama Bridge on the Kita River. **a** Before sediment removal. Photo date: March 19, 2010. **b** After sediment removal. Photo date: March 11, 2021

Fig. 2.8 Strengthened embankments (*Photos* by Akira Matsui). *Notes* The photos were taken at an open levee. **a** The photo was taken at the confluence between the Onyu River and Kita River. Photo date: February 26, 2021. **b** The photo was taken at the confluence between the Nogi River and Kita River. Photo date: October 14, 2020

Fig. 2.9 **a** Strengthened slope and **b** water level gauge (*Photos* by Akira Matsui). *Notes* **a** Photo date: October 14, 2020. **b** Photo date: February 26, 2021

Box 2.1 Traditional River Construction Method

There are many flood-related disasters due to the natural and social character-istics of rivers, and the hydraulic engineering projects constructed to protect the lives and property of residents from floods have been extremely impor-tant for a long time in Japan. Therefore, people sought the best construction methods in consideration of safety and durability and created the traditional river construction method in Japan.

However, at present, the number of traditional river construction projects is decreasing due to concerns about structure strength and durability during floods, difficulties in obtaining natural materials such as wood and stone, rising prices and a shortage of engineers.

The traditional river construction method is highly adaptable to the construc-tion site and ground changes after construction. In addition, the traditional river construction method has many advantages over modern hydraulic construction methods, such as its positive impact on the ecosystem.

The traditional river construction method uses natural materials such as wood and stone and includes foot protection work (Appendix Fig. 2.10), groin work and bed protection work; each of which has been developed over many years in Japan as a result of accumulated resourceful efforts.

When the traditional river construction method is used, the shapes of the spaces between materials are diverse, and the changes in flow velocity are various. Therefore, fish can select spaces with multiple shapes or flow velocities to move through according to their own habits. Thus, this environment can serve as a habitat to a variety of fish.

Appendix

See Fig. 2.10.

Fig. 2.10 Foot protection work. *Source* Reprinted from the Japan RiverFront research Center, http://www.rfc.or.jp/rp/files/02-03.pdf, Accessed October 30, 2021 (in Japanese)

References

Okuma T (1987) A study on the function and etymology of open levee (in Japanese). Paper of the
    Research Meeting on the Civil Engineering History in Japan, vol 7, pp 259–266. https://doi.org/
    10.11532/journalhs1981.7.259
Teramura J, Okuma T (2005) A study on evolution and role of open levees on alluvial-fan rivers
    in the Hokuriku District -from the view point of decentralization of river-engineering decision
    making (in Japanese with English Abstract). J Hist Stud Civil Eng 24:161–171. https://doi.org/
    10.11532/journalhs2004.24.161
Yoshikawa N (2014) Can paddy fields mitigate flood disaster? Possible use and technical aspects
    of the Paddy Field Dam. In: Nishikawa U, Miyashita T (eds) Social-ecological restoration in
    paddy-dominated landscapes, Ecological research monographs. Springer, Tokyo. https://doi.org/
    10.1007/978-4-431-55330-4_13

Chapter 3
Significance of Flood Disturbance

Abstract Rivers carry sediment from upstream to downstream and create natural levees and backswamps. Floods restore ecological succession and have an effect on biodiversity. Beaches on coasts are consist of earth and sand carried by a river from the mountains. When a dam is constructed in the upstream of a river, the flow in the downstream part of the river is smoothed. Consequently, the riverbed is stabilized and forested. As a process that maintains a river ecosystem, floods remove the sediment accumulated in rivers. Floods play an important role in maintaining the health of not only rivers and their ecosystems but also seas and coastal ecosystems.

Keywords Coastal ecosystem · Ecological succession · Flood disturbance · River ecosystem

3.1 What Does a Flood Carry?

Flood can destroy lives and property. However, are floods bad? No, they are not. Rivers in Japan can be understood by comparing them with the rivers across the world (Fig. 1.1). Japanese rivers are predisposed to flooding. Rivers in Japan are characterized by carrying large amounts of earth and sand from upstream, and they have created large floodplains. River sediments accumulate on both sides of river channels to create natural levees. Consequently, riverbeds become low, and wetlands form away from the river channels. These wetland areas are called backswamps (Fig. 3.1). Thus, rivers create different types of environments.

There are many organisms that need to be disturbed by floods in Japan. These organisms live in riverbanks, which are below water at the peak of flooding but are usually higher than the surface of the river. A riverbank contains stones and has no soil. If it does not rain, then it will resemble a desert without water. Therefore, the plants living on a riverbank have mechanisms that enable them to withstand droughts. Original plants on a riverbank can withstand the dryness and high temperatures of a desert, but they cannot withstand shade. There are many plants that occur on riverbanks in Japan, such as *Artemisia capillaris*, and they all possess a mechanism to withstand drought. When a riverbank is covered with tall plants and forests, many plants on a riverbank will disappear. What if floods are removed from such a system?

© The Author(s), under exclusive license to Springer Nature Singapore Pte Ltd. 2022
A. Matsui, *Wetland Development in Paddy Fields and Disaster Management*,
https://doi.org/10.1007/978-981-19-3735-4_3

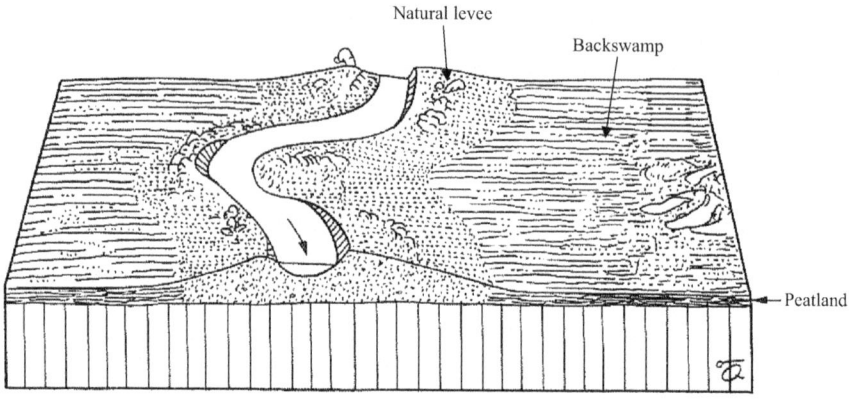

Fig. 3.1 Natural levee and backswamp. *Source* Modified from Kaizuka et al. (1995). Copyright 1995 Iwanami Shoten, Publishers

The natural levee and backswamp would disappear, and many creatures living in the river and backswamp would become extinct.

3.2 Importance of Disturbance

When dams are constructed upstream of a river, the appearance of the riverbank will change. As dams prevent floods, sediment will not flow from upstream to downstream. A river's water level does not substantially change, and many trees grow at the water's edge. As the trees increasingly grow, the riverbank becomes forested with many trees and abundant nature, but it is not in very good condition when considering the river ecosystem (Figs. 2.6a, 2.7a).

Net-spinning caddisflies build nest on the surface of river stones and set capture nets (Fig. 3.2). They prey on microscopic algae that are carried by the water flow using their capture nets. Therefore, these caddisflies inevitably need stones and gravel as a foundation, and a suitable watercourse is also needed. Thus, stone and gravel bottoms in rapids are good places for their habitats. When a large proportion of net-spinning caddisflies occurs in rapids, such a community is a tentative climax community at that location (Fig. 3.3). When net-spinning caddisflies are dominant, the macroinvertebrate assemblages are estimated to reach the climax, and species diversity decreases. Floods have helped stop ecological succession and enable systems to be restored. This phenomenon maintains the dynamic equilibrium of rivers and contributes to biodiversity.

Fig. 3.2   Net-spinning caddisfly and its capture net (Photo by Akira Matsui). *Notes* The photograph was taken in the Ooishi River in 1994. See Fig. 8.1

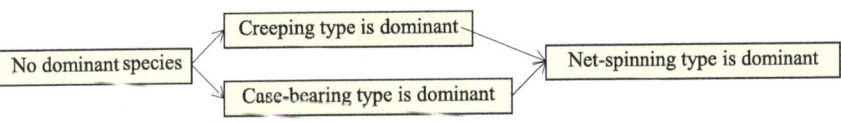

Fig. 3.3   Process of ecological succession. *Source* Modified from Tsuda (1962)

3.3   Maintaining a River Ecosystem

When dams are constructed and rivers are rehabilitated, river flows decrease. As a result, there are places where sand accumulates and becomes stationary, and the water depth becomes shallow. As a method of maintaining a river ecosystem, sediment accumulation is removed in rivers to increase flow capacity (Figs. 2.6b and 2.7b). The algae on stones at the bottom of a river serve as food for *Plecoglossus altivelis*. However, if the stones do not move, the algae will rot. In fact, floods refresh the river environment. When the flow conditions have decreased, it is important to artificially disturb the river such as through flexible dam management (see Box 8.1).

3.4   Connection Among Mountains, Rivers, Villages and Seas

Floods play an important role in maintaining the health of not only rivers but also seas. The beaches on the coast contain earth and sand, which are carried in rivers from the mountains. However, due to the construction of dams to prevent sediment-related disasters, sediment supplies are cut off, and coastlines recede. As a result, greater damage occurs in the event of a typhoon high wave or tsunami.

Forest is connected to seas. The provision of abundant brackish waters originates from forests. However, dam development and deforestation caused rapid devastation of coastal seas in the 1980s. Therefore, fishermen began planting trees in the mountains to help support the coast and organisms that live there. A tree-planting ceremony is held every year (Fig. 3.4).

Box 3.1 What are the SDGs?

SDGs are the sustainable development goals. The SDGs are defined in Appendix Table 3.1. The contents of this book relate mainly to SDGs 9 and 11. Building a resilient infrastructure must be achieved to protect humans, livelihoods and property from flood-related disasters. An effective way to achieve this goal is to create green infrastructure such as wetland development in paddy fields.

Fig. 3.4 Tree-planting ceremony. *Source* Reprinted from the NPO corporation mori wa umi no koibito, https://mori-umi.org/about/activity/activity_mori/activity_mori01/#report01, Accessed October 30, 2021 (in Japanese)

Appendix

See Table 3.1.

Table 3.1 SDGs

Goal No	Icons	SDG
Goal 1		End poverty in all its forms everywhere
Goal 2		End hunger, achieve food security and improved nutrition and promote sustainable agriculture
Goal 3		Ensure healthy lives and promote well-being for all at all ages
Goal 4		Ensure inclusive and equitable quality education and promote lifelong learning opportunities for all
Goal 5		Achieve gender equality and empower all women and girls
Goal 6		Ensure availability and sustainable management of water and sanitation for all
Goal 7		Ensure access to affordable, reliable, sustainable and modern energy for all
Goal 8		Promote sustained, inclusive and sustainable economic growth, full and productive employment and decent work for all
Goal 9		**Build resilient infrastructure, promote inclusive and sustainable industrialization and foster innovation**
Goal 10		Reduce inequality within and among countries
Goal 11		**Make cities and human settlements inclusive, safe, resilient and sustainable**
Goal 12		Ensure sustainable consumption and production patterns
Goal 13		Take urgent action to combat climate change and its impacts

(continued)

Table 3.1 (continued)

Goal No	Icons	SDG
Goal 14		Conserve and sustainably use the oceans, seas and marine resources for sustainable development
Goal 15		Protect, restore and promote sustainable use of terrestrial ecosystems, sustainably manage forests, combat desertification, and halt and reverse land degradation and halt biodiversity loss
Goal 16		Promote peaceful and inclusive societies for sustainable development, provide access to justice for all and build effective, accountable and inclusive institutions at all levels
Goal 17		Strengthen the means of implementation and revitalize the Global Partnership for Sustainable Development

Source Reprinted from the United Nations. https://www.un.org/sustainabledevelopment/news/communications-material/ Accessed October 30, 2021

References

Kaizuka S, Naruse Y, Ohta Y, Koike K (1995) Plains and Coasts in Japan New Edition Nature in Japan 4 (in Japanese). Publishers, Tokyo, Iwanami Shoten

Tsuda M (1962) Aquatic entomology (in Japanese). Hokuryukan & New Science Co., Ltd., Tokyo

Part II
Paddy Fields

Chapter 4
Characteristics of Paddy Fields

Abstract An agricultural area that occupies more than 50 ha is large in high-income country, while one that occupies less than 5 ha is large in low-income countries. Globally, production and yield quantities of paddy rice have been increased from 2001 to 2018. In 2018, the share of paddy rice production in Asian countries was large. Japan is at the high end of this production. However, the agricultural workforce has been decreasing and increasingly aging in Japan. The area and rate of abandoned agricultural land have significantly increased in Japan with the area of abandoned agricultural land of farm families being larger than that of nonfarm families, while the increasing rate of abandoned land is especially prominent for nonfarm families in Japan.

Keywords Abandoned agricultural land · Family farming · High-income country · Low-income country · Middle-income country · Paddy field

4.1 Agriculture Across the World

There has been no major change in the scale of agricultural areas globally for approximately 50 years. High middle-income countries have the largest agricultural areas. High-income countries and low middle-income countries have moderately sized agricultural areas, and low-income countries have the smallest agricultural areas (Fig. 4.1).

Of the agricultural areas in the world, agricultural land that is 50 ha or more accounts for the largest proportion. This trend is remarkable in high-income countries and high middle-income countries. On the other hand, agricultural land that is 5 ha or less accounts for the largest proportion of agricultural lands in low-income countries and low middle-income countries (Fig. 4.2).

Of the high-income countries, Australia and the United States have the largest agricultural land areas. In contrast, France, the United Kingdom, Germany and Japan have smaller agricultural areas (Fig. 4.3).

As a percentage of the agricultural areas in high-income countries, agricultural land of 50 ha or more accounts for the largest proportion of these lands in the United States, the United Kingdom, France and Germany. On the other hand, agricultural

© The Author(s), under exclusive license to Springer Nature Singapore Pte Ltd. 2022

A. Matsui, *Wetland Development in Paddy Fields and Disaster Management*,
https://doi.org/10.1007/978-981-19-3735-4_4

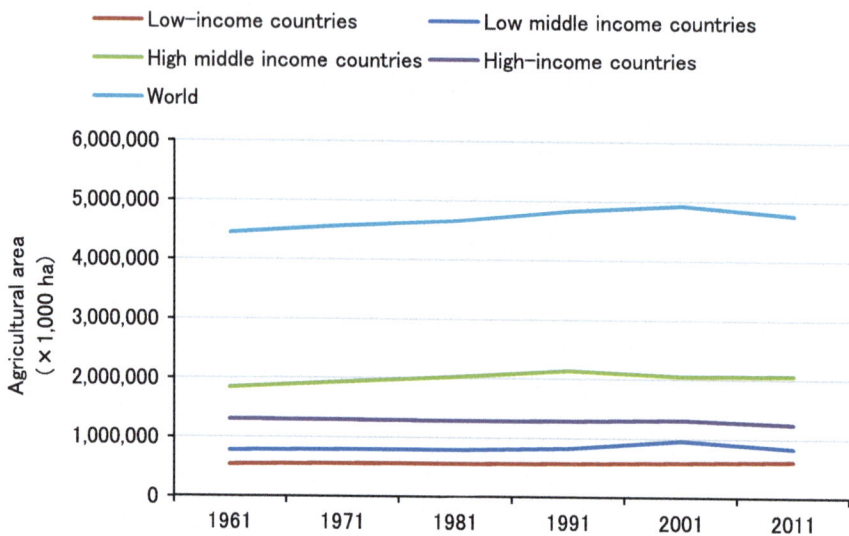

Fig. 4.1 Scale of agricultural areas in each country. *Source* Illustration based on FAO (2014)

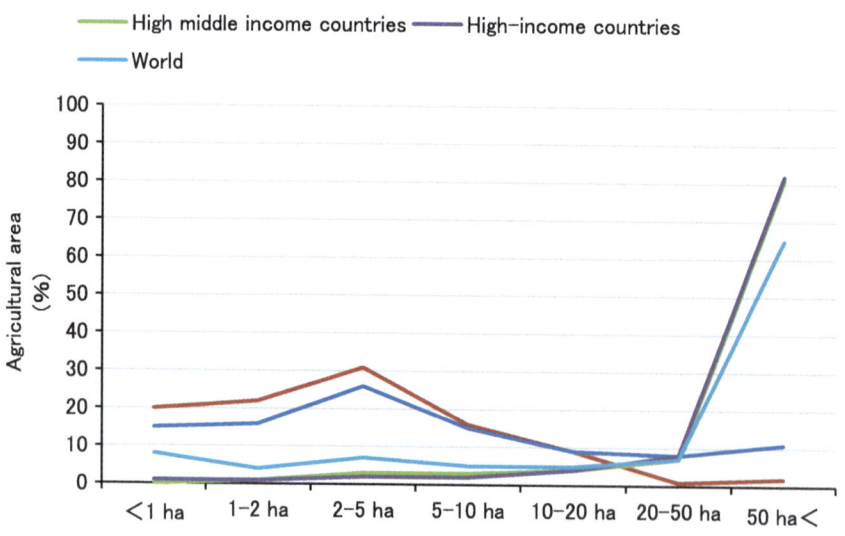

Fig. 4.2 Percentage of agricultural areas in each country (by size of agricultural land). *Source* Illustration based on FAO (2014)

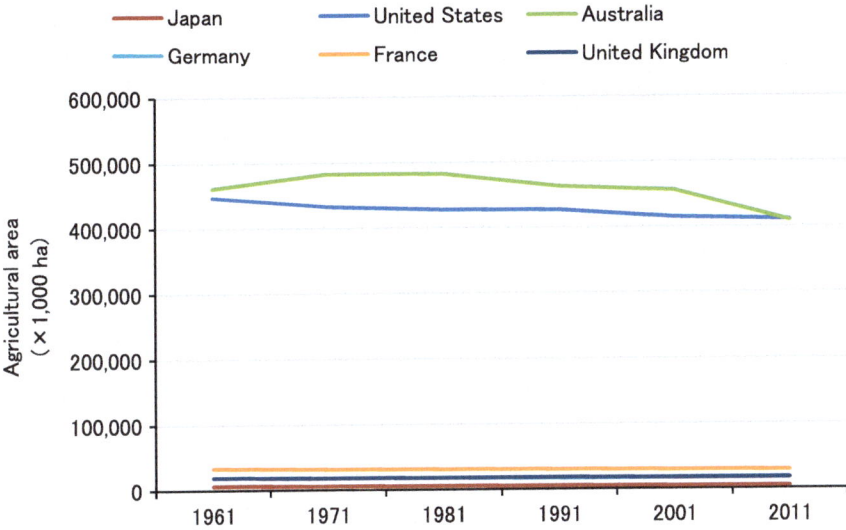

Fig. 4.3 Scale of agricultural areas in each country (high-income countries). *Source* Illustration based on FAO (2014)

land of 5 ha or less accounts for the largest proportion of the agricultural areas in Japan (Fig. 4.4).

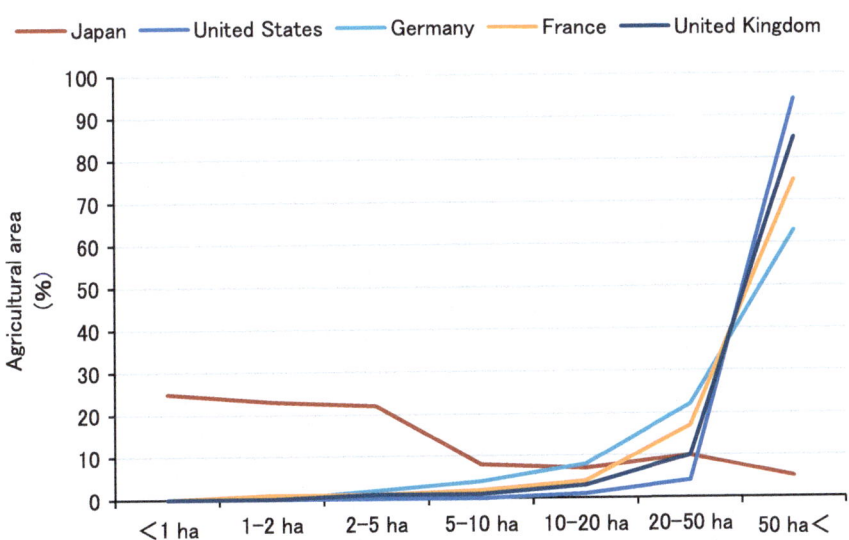

Fig. 4.4 Percentage of agricultural areas in each country (by size of agricultural land) (high-income countries). *Source* Illustration based on FAO (2014)

It should be noted that Japan's percentage of agricultural areas is similar to that of low-income countries and low middle-income countries. These countries are characterized by a focus on family farming, and family farming is central to Japanese agriculture.

The international year of family farming was 2014. Family farming includes all family-based agricultural activities, and it is linked to several areas of rural development. Both in developing and developed countries, family farming is the predominant form of agriculture in the food production sector. Family farming plays important socioeconomic, environmental and cultural roles (FAO 2021b).

Regarding the relationship between farm size and productivity, per-acre yields are unaffected by farm size (Feder 1985). Increasing farm size can be attributed to decreasing returns to scale (Cornia Giovanni 1985). The smallest effective size will be the most competitive size for farms (Gerard and John 1996). The practice of smallholding is more efficient and less environmentally degrading than industrial agriculture, which depends heavily on fossil fuels, chemical fertilizers, pesticides and herbicides (Netting Robert 1993). As mentioned above, family farming will continue to be an important form of farming throughout the world.

In Japan, the cultivated acreage per unit of family farm was 3.1 ha in 2020. On the other hand, in the United States and Australia, this value was 180.2 ha in 2005 and 3423.8 ha in 2004, respectively. In the European Union, Germany, France and the United Kingdom, the cultivate acreage per unit of family farm was 16.9 ha in 2005, 43.7 ha in 2004, 52.3 ha in 2004 and 55.4 ha in 2004, respectively (Fig. 4.5). These results mean that Australia and the United States have large farming areas, European

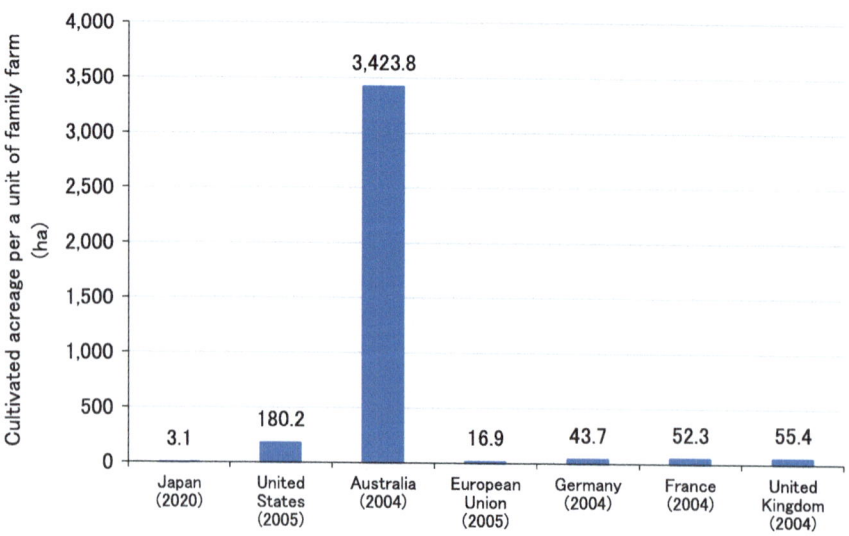

Fig. 4.5 Disparity in cultivated acreage among nations. *Note* Cultivated acreage contains paddy fields, crofts and orchards. *Source* Illustration based on Ministry of Agriculture, Forestry and Fisheries of Japan (2020)

Union nations have moderately sized farming areas, and Japan has small farming areas. The importance of the multiple functions of agriculture is not observed in the countries with large farming areas, while it is recognized in countries with small farming areas. The multiple functions of agriculture are used in Japan.

1. **Perspectives on the multiple functions of large farming areas**

   Nations with large agricultural areas, such as the United States and Cairns Group (Australia, New Zealand, Canada, etc.), do not deny the importance of environmental issues and argue that the notion of multiple functions of agricultural areas remains unclear. However, the notion of multiple functions of agricultural areas is a covert of protectionist policy, and it distorts free trade policy.

2. **Perspectives on the multiple functions of moderate farming areas**

   European Union nations have moderate farming areas with less regional integration than Japan, which primarily implements rainfed farming and grassland livestock management. The food self-sufficiency rate of European Union nations has reached over 70%. In Germany, a food self-sufficiency rate of 70% or more is common and recognized as a national right. France is a food exporter that rivals the United States, and it is already aiming to promote multiple functions of agriculture and regional revitalization rather than food security.

   European Union nations have conditional and disadvantaged regional policies that appeared early in their history. These policies provide direct income compensation to farms that are not in areas with favorable conditions, such as areas remote from cities and on mountainous slopes, and this type of agriculture is still considered meaningful.

3. **Perspectives on the multiple functions of small farming areas**

   Japan has small farming areas with strong territoriality, centered on rice farming and crop cultivation based on field rotations due to geographical and meteorological conditions. Japanese people feel uneasy because their food self-sufficiency rate is too low. They recognize the significance of the multiple functions of agriculture and are concerned that their significance will decline.

   The type of consideration and positioning a country like Japan with small farming areas will be given internationally will determine the fate of many small-farm areas including those in developing countries.

Cultivated acreage per unit of family farm has gradually increased in Japan (Fig. 4.6), and the integration percentage of cultivated acreage has increased steadily (Fig. 4.7). In addition, it has been necessary to increase farm size due to agricultural intensification in Japan. Hara (2014) identified a crisis in the fact that intense agricultural intensification was adopted at a wide scale by the Trans-Pacific Strategic Economic Partnership Agreement (abbreviation; TPP). To promote the structural reform of Japan's agriculture, a farmland middle management mechanism was created to integrate and consolidate agricultural land use at the prefectural level. According to the established mechanism, improvements to abandoned farmland

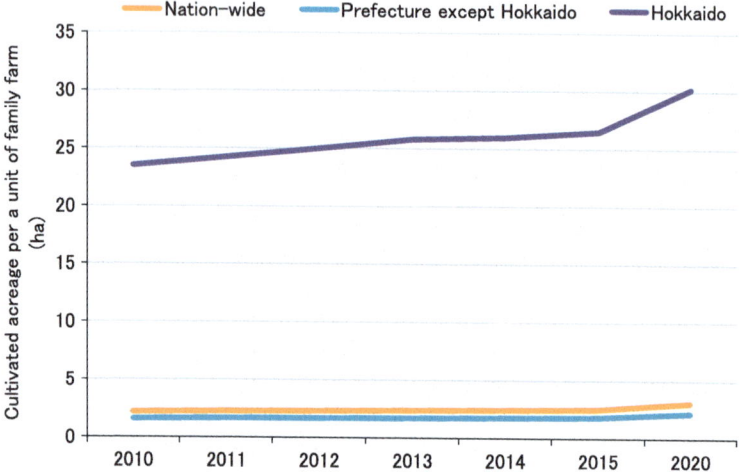

Fig. 4.6 Change in cultivated acreage per a unit of family farm in Japan. *Note* Cultivated acreage contains paddy fields, crofts and orchards. *Source* Illustration based on Ministry of Agriculture, Forestry and Fisheries of Japan (2020)

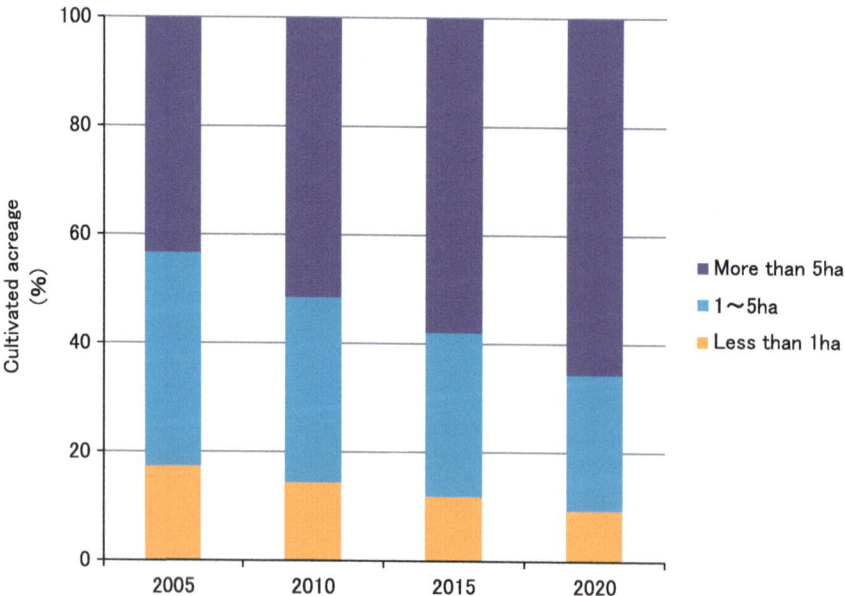

Fig. 4.7 Cultivated acreage integration percentage in Japan. *Note* Cultivated acreage contains paddy fields, crofts and orchards. *Source* Illustration based on Ministry of Agriculture, Forestry and Fisheries of Japan (2020)

strategies and strengthening measures to promote farming such as youth involvement and facilitation of investment in agricultural were implemented (Ministry of Agriculture, Forestry and Fisheries of Japan 2016).

4.2 Paddy Fields Throughout the World

The production and yield quantities of rice and paddy fields worldwide increased from 1961 to 2019 (Figs. 4.8 and 4.9). The top 10 producers from 1961 to 2019 are shown in Fig. 4.10. Japan is one of the top 8 producers. The production share of rice and paddy fields by region from 1961 to 2019 was large in Asia (Fig. 4.11). On the other hand, the production and yield quantities of rice and paddy fields in Japan decreased from 1961 to 2019 (Figs. 4.12 and 4.13).

The Asian monsoon area accounts for 14% of the world. However, this area has abundant rainfall and 54% of the world's population. In addition, this area produces 90% of the world's rice and skillfully uses water (Yamaoka 2007). Paddy fields, which are distributed widely around the world, are distributed in not only Asia but also Europe, Africa, North America and South America (Tabuchi 1999). Paddy fields exist in wet monsoon areas and arid regions globally. A rainy season and dry season exist in monsoon Asia. Alluvial lowlands cover only 1/20 of the world and exist in half of the world, but they occur in 1/6 in the Asian monsoon area. Alluvial lowland areas

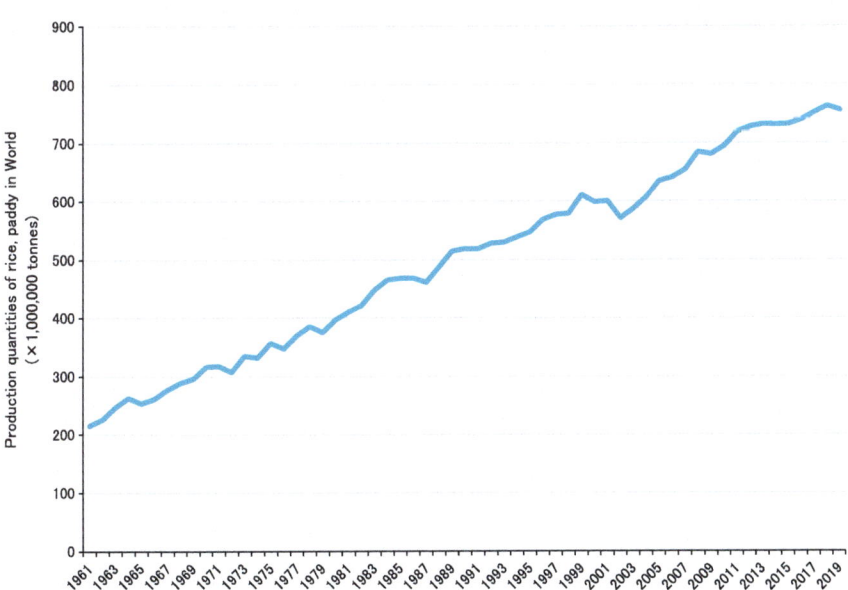

Fig. 4.8 Global production quantities of rice paddies from 1961 to 2019. *Source* Illustration based on FAO (2021a)

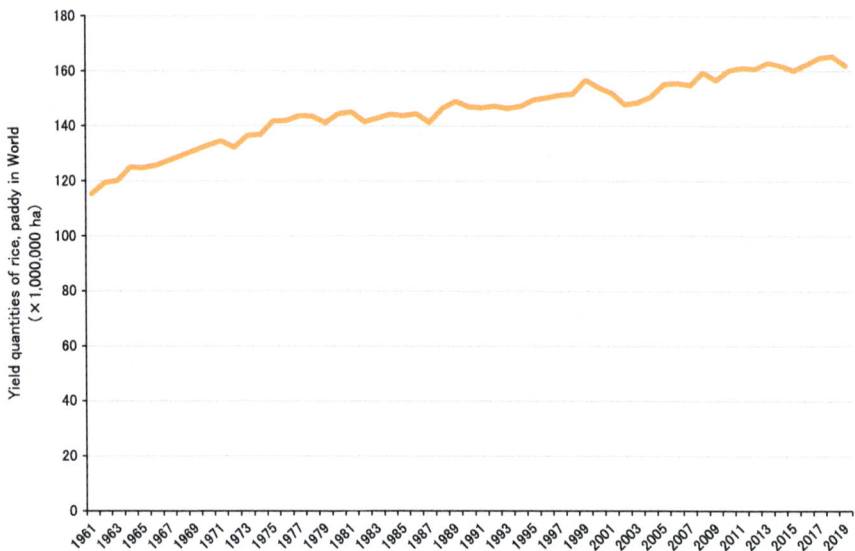

Fig. 4.9 Global yield quantities of rice paddies from 1961 to 2019. *Source* Illustration based on FAO (2021a)

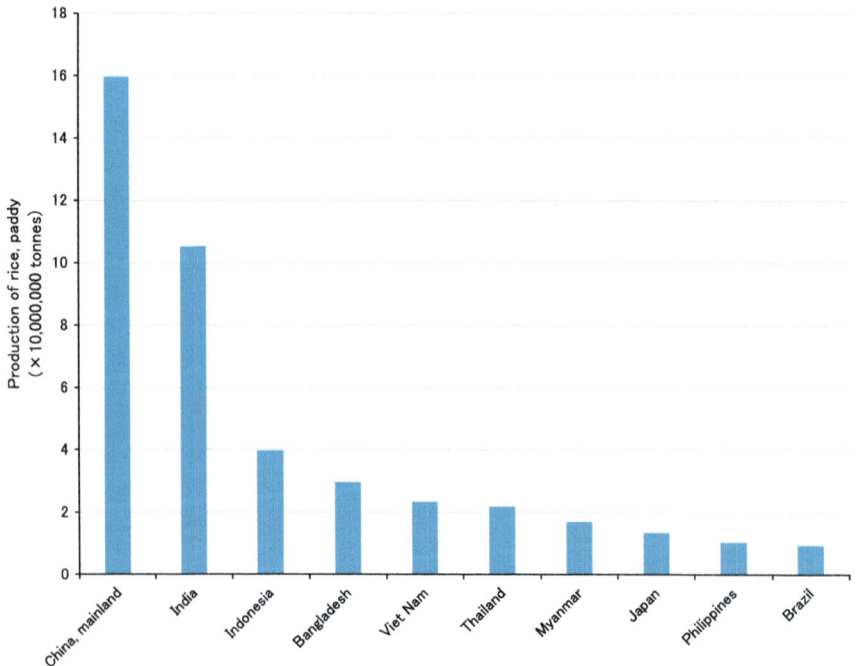

Fig. 4.10 Production of rice paddies: top 10 producers from 1961 to 2019. *Source* Illustrated based on FAO (2021a)

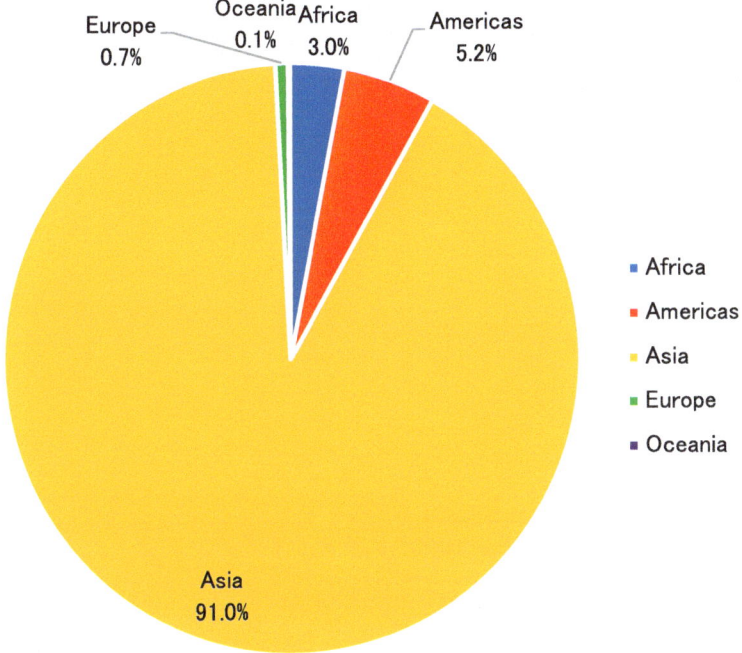

Fig. 4.11 Production share of rice paddies by region from 1961 to 2019. *Source* Illustration based on FAO (2021a)

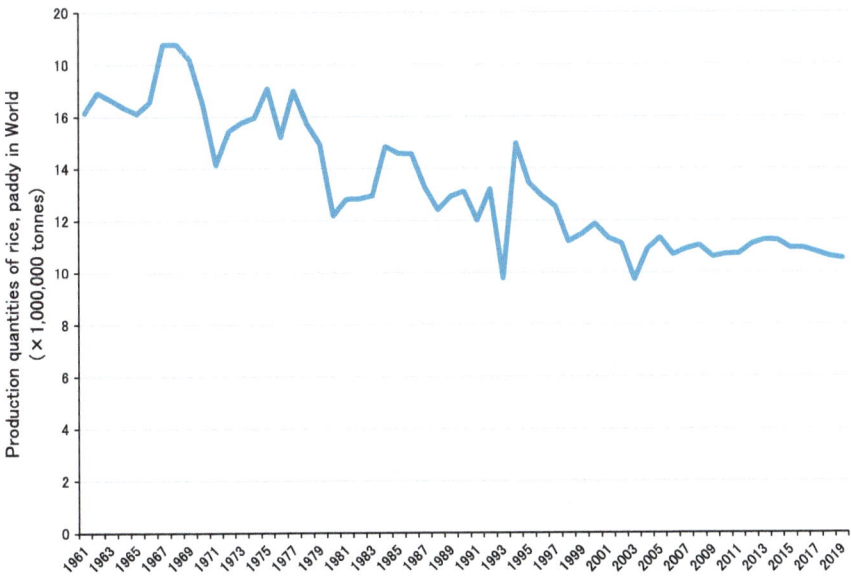

Fig. 4.12 Production quantities of rice paddies in Japan from 1961 to 2019. *Source*: Illustration based on FAO (2021a)

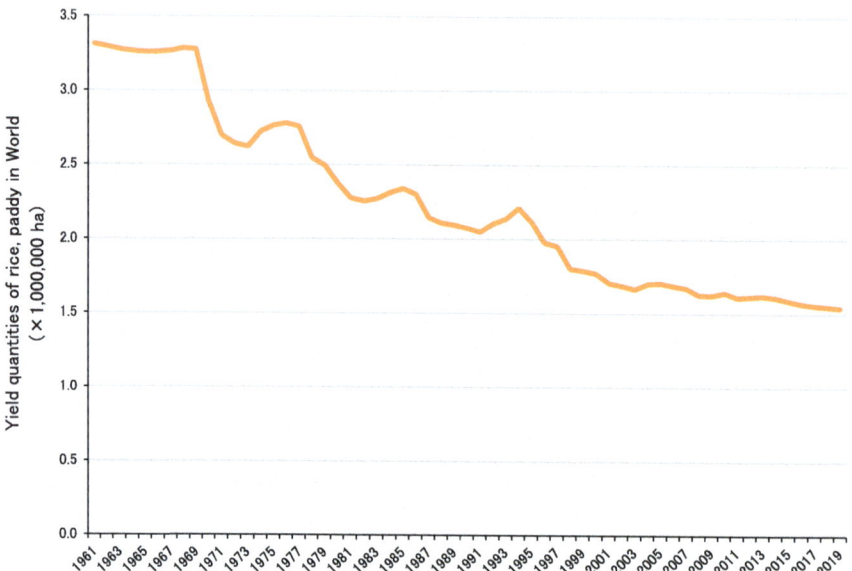

Fig. 4.13 Yield quantities of rice paddies in Japan from 1961 to 2019. *Source* Illustration based on FAO (2021a)

can be used as cultivated land in the Asian monsoon area (Kyuma 1988). Japanese rice cultivation, which occurs in monsoon Asia, is different from that in Southeast Asia. During the rainy season in Southeast Asia, there is no need for irrigation. On the other hand, Japanese rice cannot be grown without irrigation (Matsuda 1988). Rice cultivation in tropical Asia has been conducted as an adaptation to the natural environment of the area. At the same time, rice cultivation in Japan has become possible for the first time due to irrigation (Kyuma 1988).

4.3 Paddy Fields in Japan

Rice, which is the principal Japanese food, is produced in paddy fields. Paddy fields are agricultural lands that can be flooded and are surrounded by ridges (Tabuchi 1999) (Fig. 4.14). Paddy fields are composed of three layers from top to bottom: the plow layer, hardpan and subsoil layer (Fig. 4.15). The plow layer is at a depth of 10–15 cm. This layer stores nutrients and supplies them to the rice. It is the most important layer for growing rice. The hardpan restrains permeation of water and prevents machines from sinking. This layer is naturally formed by the accumulation of a solute and a mechanical run. On the other hand, it is artificially formed by rolling compaction. The subsoil layer is part of the natural soil. This layer is composed of gravel in alluvial fans and on mountain slopes, clay and sand in alluvial areas, and volcanic ash in volcanic ash areas. Groundwater is shallow in paddy fields on alluvial

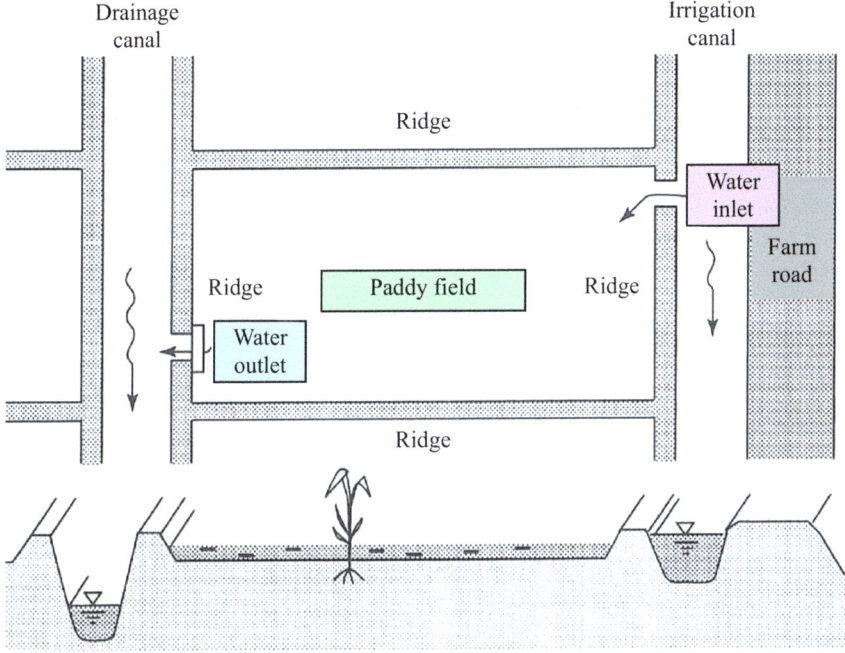

Fig. 4.14   Organization of a paddy field. *Source* Modified from Tabuchi (1999). Copyright 1999
Yamasaki Research Institute

Fig. 4.15   Soil layer in a
cross-section of a paddy
field. *Source* Modified from
Tabuchi (1999). Copyright
1999 Yamasaki Research
Institute

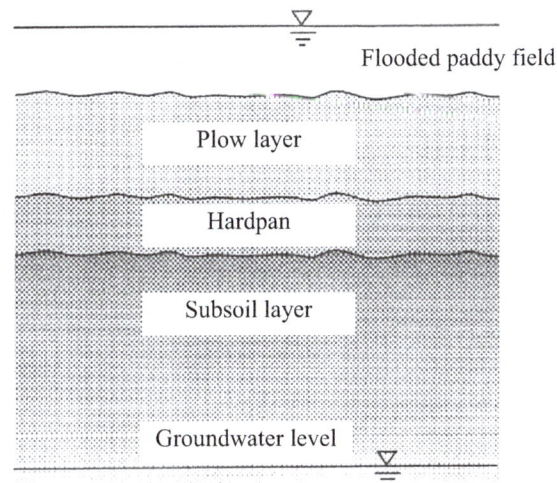

plains, and the subsoil layer is saturated with water. As a result, the subsoil layer is in a poor oxygen and reduced state. However, the groundwater level is generally quite deep in the volcanic ash area.

An irrigation canal system places the water taken from a river to paddy fields (Fig. 4.16). The water is supplied to paddy fields through a main canal, lateral canal and farm canal. An irrigation canal is generally constructed with a concrete three-sided channel to prevent water leakage and currently is a pipe channel waterway rather than an open channel so that the water can be easily accessed by twisting a faucet. As a result, it is difficult for aquatic animals to live in an irrigation canal system.

A drainage canal system returns the wastewater used in the paddy fields to the river (Fig. 4.16). The wastewater is drained to the river through the farm drain, lateral drain and main drain. There is no issue if the water in the drainage canal leaks; hence,

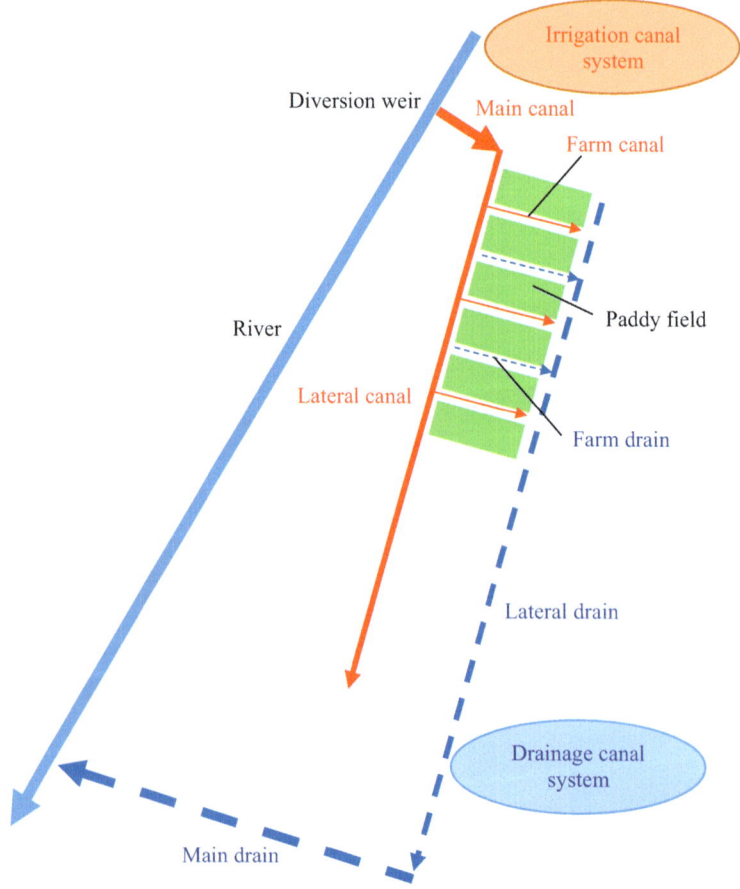

Fig. 4.16 Irrigation and drainage canal systems in a paddy field

it is not necessary to construct a concrete, three-sided line. A drainage canal is often made with concrete on a sidewall and a natural canal bed. Therefore, the drainage canal system is environmentally friendly for aquatic animals.

4.4 Present State of Paddy Fields in Japan

The agricultural workforce has been decreasing and increasingly aging in Japan (Figs. 4.17 and 4.18). The number of agricultural and private management bodies is larger than that of corporations. The former have had a pronounced tendency to decline (Fig. 4.19), and the latter have shown an increasing trend (Fig. 4.20). The area and rate of abandoned agricultural land have significantly increased (Fig. 4.21). The area of abandoned family farm agricultural land was larger than that of abandoned nonfarm family agricultural land, while the increase rate was especially prominent for the nonfarm family (Fig. 4.22). In short, these recent trends suggest that people are not concerned about abandoned agricultural farmland.

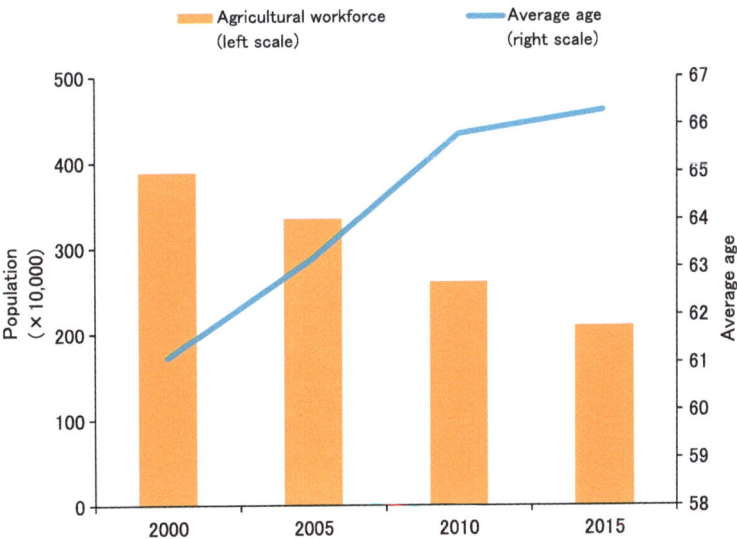

Fig. 4.17 Agricultural workforce and average age in Japan. *Source* Illustration based on Ministry of Agriculture, Forestry and Fisheries of Japan (2015)

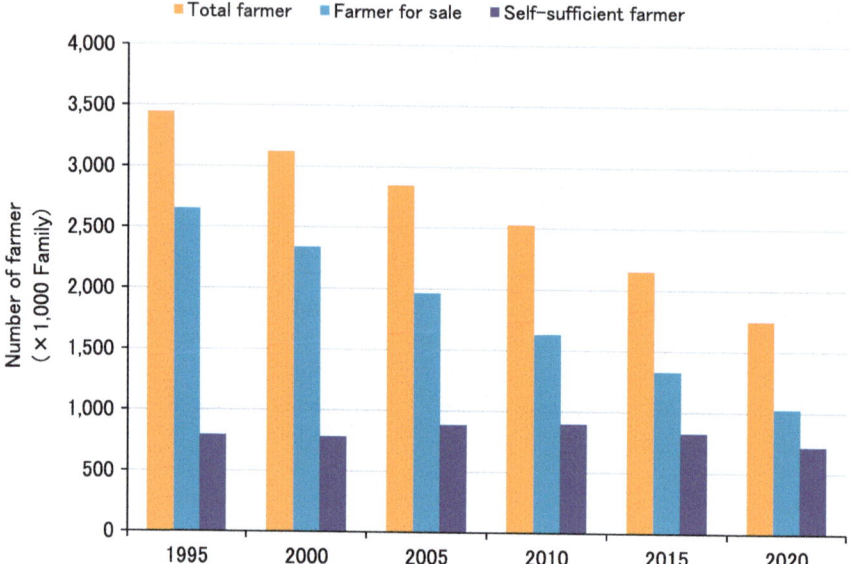

Fig. 4.18 Number of farmers in Japan. *Source* Illustration based on Ministry of Agriculture, Forestry and Fisheries of Japan (2020)

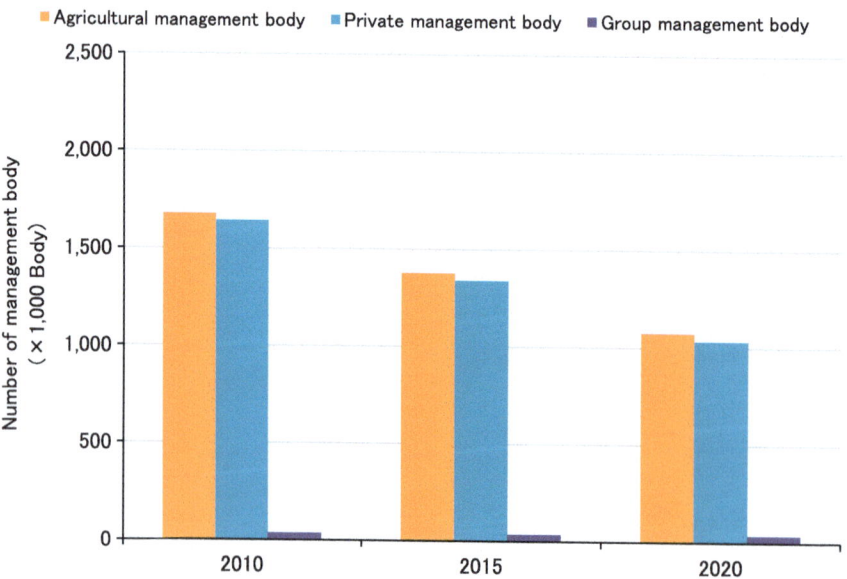

Fig. 4.19 Number of management bodies in Japan. *Source* Illustration based on Ministry of Agriculture, Forestry and Fisheries of Japan (2020)

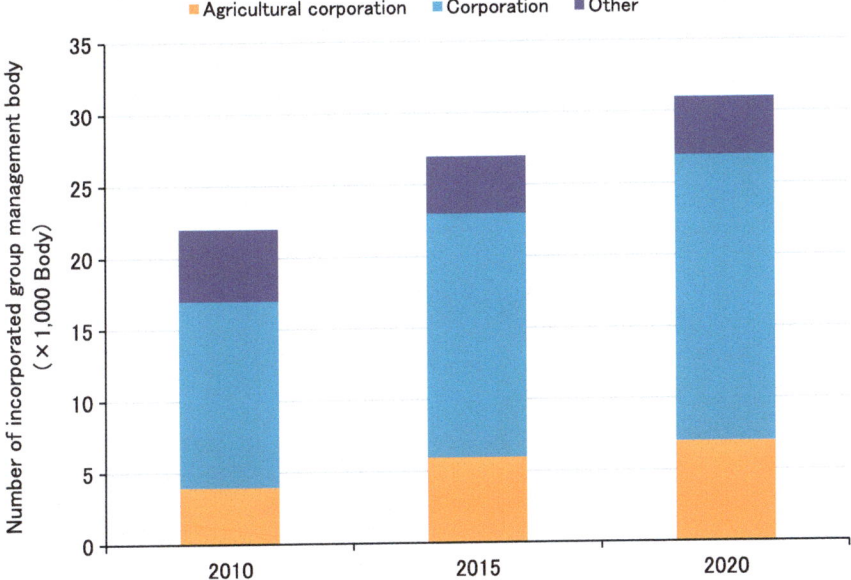

Fig. 4.20 Number of corporations in Japan. *Source* Illustration based on Ministry of Agriculture, Forestry and Fisheries of Japan (2020)

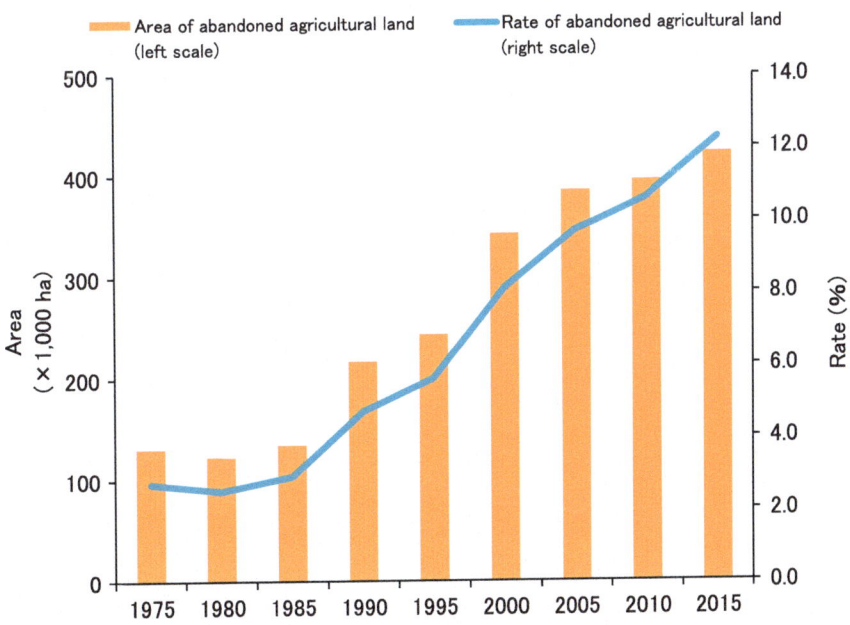

Fig. 4.21 Area and rate of abandoned agricultural land in Japan. *Source* Illustration based on Ministry of Agriculture, Forestry and Fisheries of Japan (2015)

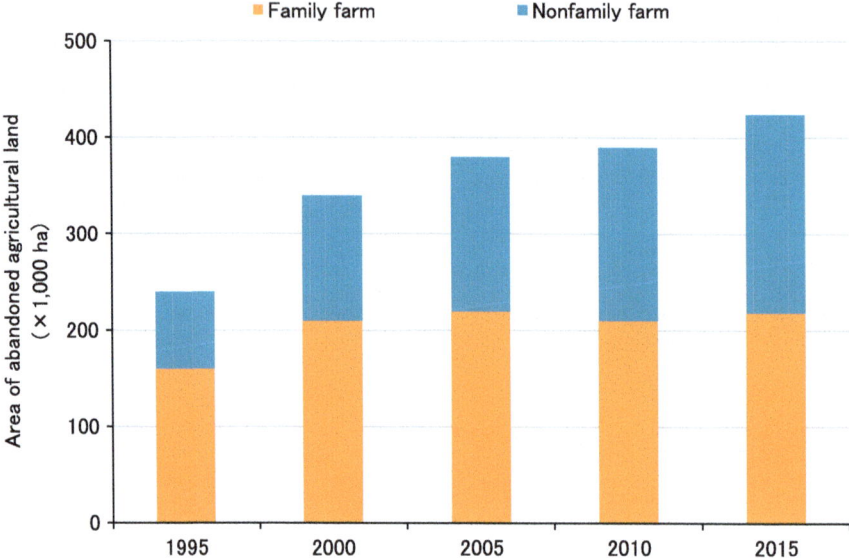

Fig. 4.22   Area of family farm and nonfamily farm abandoned agricultural lands in Japan. *Source* Illustration based on Ministry of Agriculture, Forestry and Fisheries of Japan (2015)

Box 4.1 Agriculture in Monsoon Asia

East and South Asia, where rice is mainly cultivated, is called the Asian monsoon area. A monsoon is a seasonal wind, and the predominant wind direction generally changes to the opposite direction depending on the season, which clearly alternates between the rainy season and the dry season. Thus, one of the climatic characteristics of monsoon Asia is that the change in the dry and rainy seasons is clear with precipitation concentrated in the rainy season.

Agriculturally, water scarcity in the dry season is an extremely severe condition, and there is no way to overcome it except for artificial irrigation. On the other hand, in the rainy season, it is necessary to address excess water and manage agriculture.

The conditions for establishing a rice cultivation area are a concentration of precipitation in the rainy season and the existence of vast lowlands. It seems that no conditions exist for rice cultivation to occur naturally over a large area both in terms of climate and topography in Japan.

On the other hand, Japan has a large mountainous area covered with dense forests that forms a very large catchment area. Thus, although there is no concentration of precipitation in the rainy season, there is a continuous river flow throughout a year, providing a rich water source.

Moreover, the scale of these rivers is small, and they are relatively easily controlled by humans. In addition, many of the plains created by these many

Fig. 4.23 Rice cultivation landscape. *Source* Reprinted from the National Agriculture and Food Research Organization, https://www.naro.go.jp/english/laboratory/niaes/marco/index.html, Accessed October 30, 2021

rapidly flowing rivers have an alluvial fan character with a specific slope. Therefore, if the water is blocked and divided, it becomes possible to irrigate relatively easily due to the natural gradient.

Rice cultivation (Appendix Fig. 4.23) is the best farming method for a hot and humid climate like that in Japan, and as a result, a rice-centered diet was created. On the other hand, Europe has a climate of low temperature and dryness compared to that of monsoon Asia, so livestock farming, which is similar to a crop rotation of wheat and barley, was implemented. Therefore, the eating habits of Europeans based on bread and livestock products were naturally formed.

Appendix

See Fig. 4.23.

References

Cornia Giovanni A (1985) Farm size, land yields and the agricultural production function: an analysis for fifteen developing countries. World Dev 13:513–534. https://doi.org/10.1016/0305-750X(85)90054-3

Gerard D, John I (1996) Small farms and sustainable development: is small more sustainable? J Agric Appl Econ 28:73–83. https://doi.org/10.1017/S1074070800009470

FAO (2014) The State of Food and Agriculture Innovation in family farming. http://www.fao.org/3/i4040e/i4040e.pdf. Accessed 30 Oct 2021

FAO (2021a) FAOSTAT crops. http://www.fao.org/faostat/en/#data/QC/visualize. Accessed 21 June 2021

FAO (2021b) United Nations decade of FAMILY FARMING 2019–2028. http://www.fao.org/family-farming-decade/en/. Accessed 19 June 2021

Feder G (1985) The relation between farm size and farm productivity: the role of family labor, supervision and credit constraints. J Dev Econ 18:297–313. https://doi.org/10.1016/0304-3878(85)90059-8

Hara K (2014) International year of family farming 2014—the worth of family farming is called into question now. Norin Kinyu 67:53–59 (in Japanese)

Kyuma K (1988) System as paddy field. In: Sofue T (ed) The Asian society judged from rice. NHK Publishing Co., Ltd., Tokyo, pp 41–55 (in Japanese)

Matsuda T (1988) The village which makes rice. In: Sofue T (ed) The Asian society judged from rice. NHK Publishing Co., Ltd., Tokyo, pp 56–68 (in Japanese)

Ministry of Agriculture, Forestry and Fisheries of Japan (2015) Report on results of Longitudinal Census of agriculture in Japan. https://www.maff.go.jp/j/tokei/census/afc2015/280624.html. Accessed 19 June 2021 (in Japanese)

Ministry of Agriculture, Forestry and Fisheries of Japan (2016) About a farmland middle management mechanism. http://www.maff.go.jp/j/keiei/koukai/kikou/. Accessed 19 June 2021 (in Japanese)

Ministry of Agriculture, Forestry and Fisheries of Japan (2020) Report on results of Longitudinal Census of agriculture in Japan. https://www.maff.go.jp/j/tokei/census/afc/2020/. Accessed 30 Oct 2021 (in Japanese)

Netting Robert M (1993) Smallholders, householders: farm families and the ecology of intensive, sustainable agriculture. Stanford University Press, California

Tabuchi T (1999) Paddy fields in the World. Yamasaki Research Institute, Tokyo (in Japanese)

Yamaoka K (2007) Sustainable agriculture in the Asian monsoon region and creation of water environment and culture—water network system of increasing importance. Keynote speech at 1st Asia-Pacific Water Summit in OITA. http://www.maff.go.jp/kyusyu/nn/info/2007120201.pdf. Accessed 20 June 2021 (in Japanese)

Chapter 5
Multiple Functions of Paddy Fields

Abstract Paddy fields provide a variety of functions. Food production is a support function that ensures a reliable and sustainable food supply for the future. Paddy fields are surrounded by ridges that serve the function of preventing and mitigating flood runoff by temporarily storing rainwater. The collapse of sloping farmland is prevented by identifying and repairing potential collapses at an early stage through agricultural production activities. Most irrigation water on paddy fields penetrates underground, and some is returned to rivers through drainage channels. Organic matter is mineralized by microorganismal activity, and the organic matter content in the water decreases in paddy fields. Properly managed farmland, such as cultivated vegetation that absorbs carbon dioxide through photosynthesis, has the ability to purify the air. Additional services such as biodiversity conservation, land space conservation, community promotion, preservation of traditional culture, human welfare and environmental education are included in the multiple functions of paddy fields.

Keywords Atmosphere control · Flood control · Food production · Groundwater recharge · Multiple functions · River flow stability · Sediment collapse prevention · Sediment runoff prevention · Water purification

5.1 Introduction

Paddy fields in Japan are mainly located on low-lying plains, while rice terraces are located on mountains. The functions of paddy fields involve not only food production but also biodiversity conservation, culture provision and landscape-related services. Generally, paddy fields located on low-lying plains are areas of food production; in contrast, rice terraces situated on mountains are largely areas providing biodiversity conservation, culture and landscape-related services. That is, Japanese rural areas are recognized as having multiple agricultural functions.

The problems that exist in both low-lying plain and mountainous regions are as follows: Conventional paddy farmlands are consolidated on low-lying plains to increase farming efficiency by improving the drainage conditions of paddy fields and independently creating irrigation and drainage canals, resulting in negative impacts on biodiversity in rural areas (Fujioka and Lane 1997; Katano et al. 2001; Lane and

© The Author(s), under exclusive license to Springer Nature Singapore Pte Ltd. 2022   49
A. Matsui, *Wetland Development in Paddy Fields and Disaster Management*,
https://doi.org/10.1007/978-981-19-3735-4_5

Fujioka 1998; Matsui and Satoh 2004a, b). In addition, mountainous regions have experienced depopulation; thus, maintaining rice terraces has been difficult.

Generally, the multiple functions of agriculture are food production, flood control, sediment collapse prevention, sediment runoff prevention, river flow stability and groundwater recharge, water purification, atmospheric control, etc.

5.2   Food Production

Food production is a support function that provides people with a reliable and sustainable food supply for the future. When international food production and distribution are hindered by global social and economic activities or untimely natural phenomena, having fresh food securely and safely produced and supplied is the most important factor in people's lives.

5.3   Flood Control

Paddy fields are surrounded by ridges and prevent and mitigate flood runoff by temporarily storing rainwater (Fig. 5.1). The greater the paddy field area is, the lower the peak flow rate and direct runoff rate in a basin. Urbanization and abandoned farmlands will increase the peak discharge during floods.

Harrowing is a process by which soil is shredded, and then, it can be mixed with water in the paddy fields (Fig. 5.2). As a result, water can be stored in paddy fields by harrowing.

Fig. 5.1 Paddy field flood control. *Source* Modified from the Ministry of Agriculture, Forestry and Fisheries of Japan, https://www.maff.go.jp/j/nousin/noukan/nougyo_kinou/pdf/adult_zentai. pdf, Accessed October 31, 2021 (in Japanese)

Fig. 5.2   Roll of harrowing in paddy fields. *Source* Modified from the Ministry of Agriculture, Forestry and Fisheries of Japan, https://www.maff.go.jp/j/nousin/noukan/nougyo_kinou/pdf/adult_zentai.pdf, Accessed October 31, 2021 (in Japanese)

Fig. 5.3   Paddy field dam. *Source* Modified from the Ministry of Agriculture, Forestry and Fisheries of Japan, https://www.maff.go.jp/j/nousin/noukan/nougyo_kinou/pdf/adult_zentai.pdf, Accessed October 31, 2021 (in Japanese)

A paddy field dam (Figs. 5.3 and 2.4) is a paddy field that acts as a dam by installing a weir board with a hole smaller than the drain pipe at the drain outlet of the paddy fields, suppressing the outflow of water. Implementing this approach in many paddy fields results in effective water storage in the paddy fields during heavy rains, reducing floods in the downstream areas.

5.4   Sediment Collapse Prevention

Preventing the total collapse of sloping farmland can be achieved by identifying and repairing farmland collapses at an early stage during agricultural production. Agricultural production enables the slow infiltration of irrigation water and maintains the groundwater level in a stable state with a cultivated board that is formed under the soil layer in paddy fields.

Abandoning cultivation causes cracks in the cultivated board. Then, rapid underground infiltration occurs during heavy rains, groundwater levels rise, and landslides

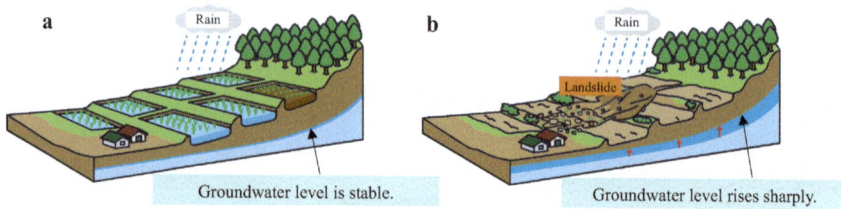

Fig. 5.4 Sediment collapse prevention. *Note* **a** Cultivation continues and **b** farming is abandoned. *Source* Modified from the Ministry of Agriculture, Forestry and Fisheries of Japan, https://www. maff.go.jp/j/nousin/noukan/nougyo_kinou/pdf/adult_zentai.pdf, Accessed October 31, 2021 (in Japanese)

occur. In addition, if farmlands are not monitored and managed, then small-scale collapses will be overlooked, and large collapses will easily occur in succession (Fig. 5.4).

5.5   Sediment Runoff Prevention

Soil erosion is a phenomenon in which soil flows out or scatters due to the action of rainwater or wind. The former is called water erosion, and the latter is called wind erosion. In comparison to bare land or wasteland with no vegetation, agricultural land experiences less soil erosion, serving as soil erosion prevention function. In paddy fields, rainfall does not act on the soil surface under flooded conditions; thus as the soil surface is flat, even in sloping areas, the soil erosion prevention function is similar to that on land that becomes wasteland due to abandoned cultivation. The leaves and stems of crops in fields weaken the impact of raindrops and the ability of surface runoff to strip the soil. Although there are differences depending on the vegetation coverage and crops, a certain degree of sediment runoff prevention function is recognized. Paddy fields have a high soil coverage and exhibit a good sediment runoff prevention function (Fig. 5.5).

5.6   River Flow Stability and Groundwater Recharge

Most irrigation water in paddy fields penetrates underground, and some is returned to rivers through drainage canals. In addition, part of the underground seepage water flows out downstream and is returned to the river, and part of it penetrates deeply as groundwater.

   When this water is returned to the river, it stays in the agricultural area, decreasing the fluctuations in the amount of river flow. This process contributes to the stability

Fig. 5.5 Sediment runoff preventions. *Note* **a** Rice is solidifying the soil and **b** farmers are solidifying the soil. *Source* Modified from the Ministry of Agriculture, Forestry and Fisheries of Japan, https://www.maff.go.jp/j/nousin/noukan/nougyo_kinou/pdf/adult_zentai.pdf, Accessed October 31, 2021 (in Japanese)

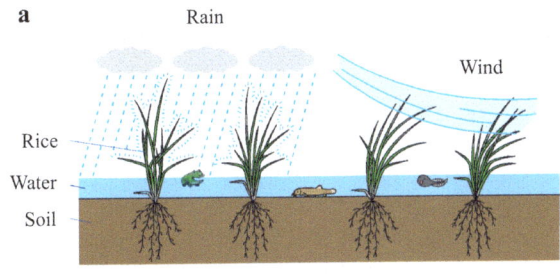

of the flow as a water source of the downstream river, and this water is reused as urban water. In addition, the water that has permeated deeply recharges the shallow and deep groundwater in the basin and is pumped up again as part of the water supply (Fig. 5.6).

Fig. 5.6 River flow stability and groundwater recharge. *Source* Modified from the Ministry of Agriculture, Forestry and Fisheries of Japan, https://www.maff.go.jp/j/nousin/noukan/nougyo_kinou/pdf/adult_zentai.pdf, Accessed October 31, 2021 (in Japanese)

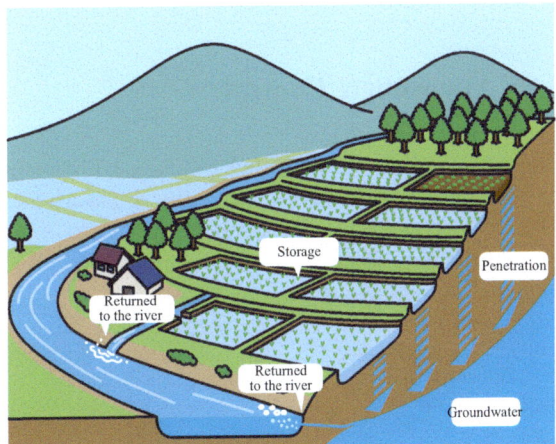

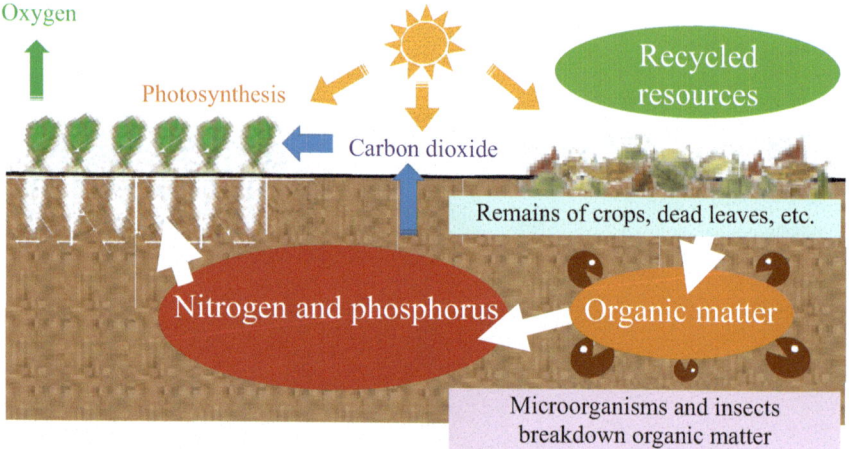

Fig. 5.7 Water purification. *Source* Modified from the Ministry of Agriculture, Forestry and Fisheries of Japan, https://www.maff.go.jp/j/nousin/noukan/nougyo_kinou/pdf/adult_zentai.pdf, Accessed October 31, 2021 (in Japanese)

5.7 Water Purification

Organic matter is mineralized by microorganismal activity in paddy fields, and the organic matter content in the water decreases. Bacteria and other microorganisms in the soil of the paddy fields decompose compost (organic matter) from livestock excrement and vegetable waste and change it into a form that is easily used by crops for nutrients (Fig. 5.7).

5.8 Atmosphere Control

Properly managed farmland, such as cultivated vegetation that absorbs carbon dioxide through photosynthesis, has the ability to absorb air pollutants and purify the air. The air is cooled by the evaporation on the water surface of the paddy fields and the transpiration of crops. This cool air has the effect of suppressing temperature increases in the surrounding city area (Fig. 5.8).

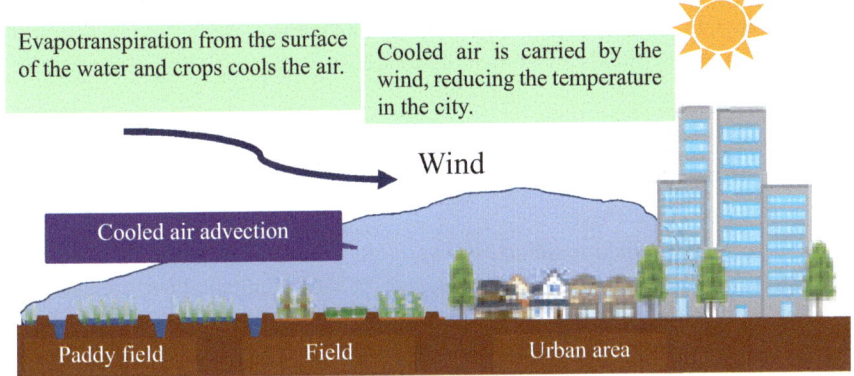

Fig. 5.8 Atmosphere control. *Source* Modified from the Ministry of Agriculture, Forestry and Fisheries of Japan, https://www.maff.go.jp/j/nousin/noukan/nougyo_kinou/pdf/adult_zentai.pdf, Accessed October 31, 2021 (in Japanese)

5.9 Biodiversity Conservation[1]

In farmland ecosystems, human actions (disturbance) temporarily displace biota; however, if there is an area to plant seeds around these farmland ecosystems, then the original biota can be restored, or a new biota will be created. In particular, it is important that paddy fields in Japan function in connection with rivers through irrigation canals. Thus, artificial restoration actions may result in more diverse biota than that which naturally occurs.

5.10 Land Conservation[2]

The dynamic conservation (maintaining/securing) of high-quality farmland guarantees sustainable agricultural production activities. Conserving farmland is also an extremely important function that ensures people that fresh food will be stably and safely produced (supply) in the future.

Furthermore, the agricultural production areas maintained in rural areas near cities provide green spaces for local communities in addition to providing various other functions. These areas are utilized as disaster prevention/evacuation spaces. They are also important to the formation of local communities due to the coexistence of a city and rural areas.

Conservation of the original Japanese landscape against the backdrop of satoyama is also important in terms of emphasizing the weight and importance of history and

---

[1] *Source* Modified from Science Council of Japan (2001).

[2] *Source* Modified from Science Council of Japan (2001).

culture to the Japanese people or the preservation of the heart and soul of Japan. These landscapes are important, and the history and culture constructed by ancestors in the terraced paddy fields are something to see.

5.11   Community Promotion[3]

Existing prefectures and municipalities were often designated based on the natural topography and the agrarian society. Therefore, in such municipalities, social capital such as farm roads, irrigation and drainage facilities and water source recharge forests that have been developed for many years accumulates and contributes to the maintenance and development of the entire community beyond agriculture.

The characteristics of a region are based on the rural society and are expressed as the identity of a local community even in the presence of industrialized urbanization. Similar cases can be seen all over the country, and urban development itself is based on the agricultural society.

5.12   Preservation of Traditional Culture[4]

The technology used in agriculture and forestry is based on various insights and folklore from ancestors and attachment to the community. Local communities have always shown a deep interest in preservation of traditional culture. The legacy of historical industries and traditional cultures that are closely related to agriculture is highlighted at municipal commemorative events. From the standpoint of regional development, traditional culture will be permanently exhibited in museums and used for names and nicknames.

5.13   Human Welfare[5]

The importance of providing human well-being functions in regions with mainly agriculture and forestry is increasing due to the stress caused by the complex issues surrounding modern political and economic society. Clean water and air, such as at oxygen bars, and quiet music in cities provide comfort and healing and have become the driving force of therapy. The welfare function of health and rest provided by the four seasons has been fully demonstrated in rural areas.

---

[3] *Source* Modified from Science Council of Japan (2001).

[4] *Source* Modified from Science Council of Japan (2001).

[5] *Source* Modified from Science Council of Japan (2001).

The healing power of nature can be seen in the fact that sea bathing was introduced into modern society for medical purposes. Seaside recreation centers were opened in various places, and many sanatoriums were constructed in places surrounded by nature. In particular, it has been medically confirmed that touching the soil, such as in horticulture and raising plants, and animals plays a role in functional recovery rehabilitation. In recent years, forest bathing has become popular at green resorts. The agriculture and forestry systems that are developed in a space where lush natural ecosystems are maintained provide an infinite place of peace for modern people who are living with urban-related tension.

5.14 Environmental Education[6]

In developed countries, satiety is increasing, and lifestyle-related issues such as increased abandonment of agricultural lands and lack of food are occurring. Underlying these issues, food is commercialized, and people are unaware that processed food depends on agricultural products (or fishery and natural harvests). As a result, gratitude and awe for life are lost, and it can be considered that society as a whole is heading in a negative direction.

With regard to such problems, by experiencing nurturing through agriculture and forests, we reaffirm the dignity of life. Experiencing nature through farming, exploring mountain and fishing nurtures human sensibilities and emotions. By experiencing the continuous cultivation of animals and plants through agriculture, we can gain an understanding of the importance of life and awe and gratitude for nature. As above, agriculture and forests provide important functions related to emotional development and environmental education.

Box 5.1 100 Selections of Rice Terraces in Japan

In July 1999, the Ministry of Agriculture, Forestry and Fisheries certified rice terraces in 117 municipalities and 134 districts throughout Japan as the '100 selections of rice terraces in Japan.' Rice terraces are rice fields created on slopes by ancestors who carved out areas on mountains and in valleys and piled up stone walls. These rice terraces are the result of ancestor wisdom and difficult experiences.

The terraced rice fields are the original scenery of Japan, where rice terraces of various sizes are gathered in a staircase pattern, and at the same time, they are examples of national land conservation and environmental conservation. The maintenance of rice terraces is difficult, but this beautiful landscape should be protected forever (Appendix Fig. 5.9).

---

[6] *Source* Modified from Science Council of Japan (2001).

Fig. 5.9 Shiroyone
Senmaida in Wajima City,
Ishikawa Prefecture, Japan
(*Photo* by Akira Matsui).
Note Photo date: June 2,
2013. *Source* Reprinted from
the 100 Selections of Rice
Terraces in Japan, http://
www.tamano.or.jp/usr/sum
iyosi/sub10.html, Accessed
October 31, 2021 (in
Japanese)

Appendix

See Fig. 5.9.

References

Fujioka M, Lane SJ (1997) The impact of changing irrigation practices in rice fields on frog popu-
lations of the Kanto Plain, central Japan. Ecol Res 12:101–108. https://doi.org/10.1007/BF0252
3615

Katano O, Hosoya K, Iguchi K, Aonuma Y (2001) Comparison of fish fauna among three types of
rice fields in the Chikuma River basin, Japan (in Japanese with English Abstract). Jpn J Ichthyol
48:19–25. https://doi.org/10.11369/jji1950.48.19

Lane SJ, Fujioka M (1998) The impact of changes in irrigation practices on the distribution of
foraging egrets and herons (Ardeidae) in the rice fields of central Japan. Biol Cons 83:221–230.
https://doi.org/10.1016/S0006-3207(97)00054-2

Matsui A, Satoh M (2004a) Distribution of aquatic animals in the drainage systems created by paddy
farmland consolidation in Shimodate City, Ibaraki Prefecture, Japan (in Japanese with English
Abstract). Jpn J Conserv Ecol 9:153–163. https://doi.org/10.18960/hozen.9.2_153

Matsui A, Satoh M (2004b) A proposal for fish habitat improvement based on the analysis of fish
distribution in the irrigation and drainage systems of a consolidated paddy field (in Japanese with
English Abstract). Ecol Civ Eng 7:25–36. https://doi.org/10.3825/ece.7.25

Science Council of Japan (2001) Evaluation of multifaceted functions of agriculture and forests
related to the global environment and human life (report) (in Japanese)

Part III
Case Study of a River

Chapter 6
Flood Survey

Abstracts Discharge and water quality were measured during times of normal flows and flooding for eleven years, from 2006 to 2016, in the Kochi River, Wakasa Town, Fukui Prefecture, Japan. The results revealed that both inorganic matter (soil particles) and organic matter (phytoplankton, nitrogen compounds and phosphorus compounds) flowed in large amounts during flooding compared to that during normal flows. Organic and inorganic matter generally flowed through the upper, middle and lower stream areas and was finally supplied to Obama Bay, an area of enclosed coastal seas. The amount of organic and inorganic matter flowing was much larger during floods than during stable periods. Floods endanger the lives and properties of humans living in basins; however, floods also play an important role in reversing transitions in river ecosystems and supplying organic and inorganic matter to marine ecosystems.

Keywords Enclosed coastal seas · Flooding · Inorganic matter · Kita River · Marine ecosystem · Normal flow · Obama Bay · Organic matter · River ecosystem

6.1 Introduction

The Kita River basin has experienced numerous floods and droughts, and the Kochigawa Dam (35°25′57″N, 135°53′43″E) was constructed to address these problems (Figs. 6.1 and 6.2). The Kochigawa Dam has a catchment area of 14.5 km^2, a flooded area of 0.37 km^2, a bank height of 77.5 m, a bank-top length of 202.3 m, a flood maximum water level (surcharge water level) of 197.7 m in elevation, a constant full water level of 190.5 m in elevation, a minimum water level of 162.7 m in elevation and a total water storage capacity of 8 million m^3. It was constructed in 2012 to provide flood control and maintain the normal function of running water, specific irrigation water, tap water and industrial water, and it was completed in 2019 (Kochigawa Dam Construction Office, Reinan Promotion Bureau, Fukui Prefecture 2010).

This chapter is a revised version of Matsui (2018). Copyright 2018 Ecology and Civil Engineering Society, https://doi.org/10.3825/ece.20.237, Accessed November 27, 2021.

© The Author(s), under exclusive license to Springer Nature Singapore Pte Ltd. 2022
A. Matsui, *Wetland Development in Paddy Fields and Disaster Management*,
https://doi.org/10.1007/978-981-19-3735-4_6

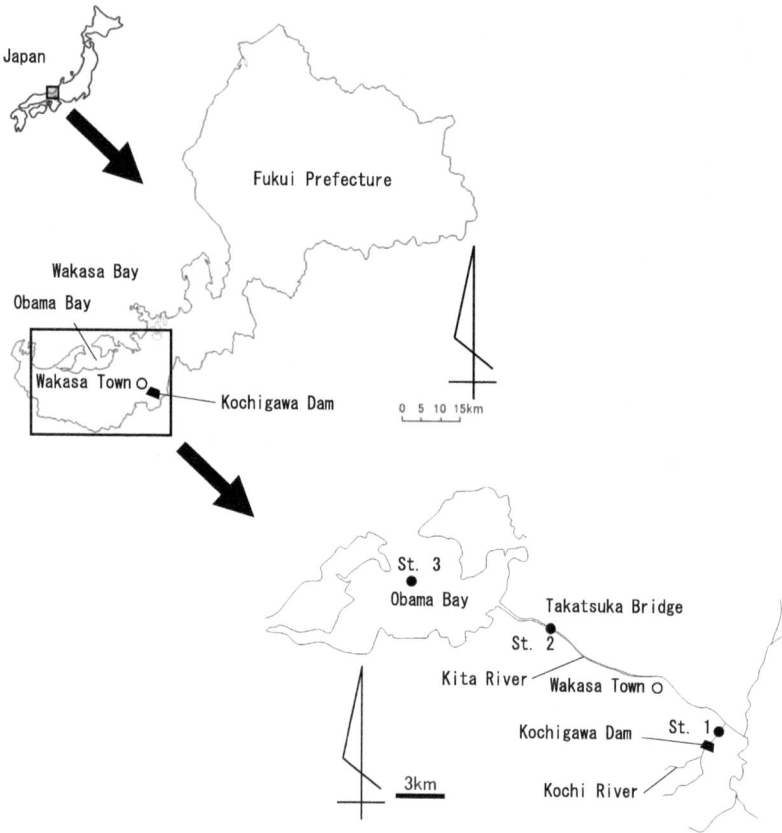

Fig. 6.1 Location of survey sites. *Source* Reprinted from Matsui (2018). Copyright 2018 Ecology and Civil Engineering Society

This dam was constructed in a way that did not affect the aquatic environment in the basin. In general, water level control based on flow rate observations is performed for flood control, which is the most important purpose of dams. Water quality management is carried out during normal flow and is rarely carried out during flooding due to the risk of disasters. Data on the concentrations of floating sediment during flooding in natural rivers are necessary for developing measures against prolonged turbid water associated with dam construction, but the data are not always sufficiently collected and organized (Sumi 1997). Furthermore, there are very few cases of surveys from rivers to sea areas that also contain research on organic matter.

In this survey, the river characteristics of the Kochi River where the Kochigawa Dam is constructed were determined by observing and comparing the flow rate and water quality during normal flows and flooding for 11 years from 2006 to 2016. The purpose of this survey is to contribute to the construction of future dam projects.

a

b

Fig. 6.2 Kochigawa Dam. *Notes* The photos were taken on December 21, 2019. **a** Long-distance view and **b** short-distance view. *Source* Reprinted from the Kochigawa Dam/Otsuro Dam Integrated Management Office, http://info.pref.fukui.lg.jp/otsuro/, Accessed October 31, 2021 (in Japanese)

6.2 Methods

St. 1 was set in the Kochi River, 1.4 km downstream from the Kochigawa Dam; St. 2 was set in the Kita River (Takatsuka Bridge), 15.9 km downstream from St. 1; and St. 3 was set in the central part of Obama Bay, 3.7 km downstream from St. 2 (Fig. 6.1). St. 1 was located in the upper reaches of the Kita River basin in a forested area. St. 2 was located in the middle and lower reaches, and the paddy field area was widespread.

For St. 1, for the normal flow survey, flow observations and water quality analyses were conducted four times a year (four seasons) from 2006 to 2016, for a total of 44 observations and analyses. For the flow observations, the water depth was measured at equal intervals with respect to the river width, and at the same time, the flow velocity at a 60% water depth was measured using an electromagnetic current meter (EM CURRENT METER ES7603 manufactured by Yokogawa Electronics Co., Ltd.). For the water quality analysis, water was sampled in a plastic bottle at the center of the stream and taken back to the laboratory for analysis. One sample for each survey, a total of 44 samples, was used for the water quality analysis. The analyzed factors were suspended solids and turbidity (soil turbidity index), chlorophyll a, total organic carbon, total nitrogen, total phosphorus and chemical oxygen demand (water quality eutrophication index).

The suspended solids, turbidity, chlorophyll a, total organic carbon, total nitrogen, total phosphorus and chemical oxygen demand were analyzed according to Environmental Agency Notification 59 Appendix Table 9, JIS K 0101 9, river water quality test method (draft), JIS K 0102 22.1, JIS K 0102 45.6, JIS K 0102 46.3.4 and JIS K 0102 17, respectively. Figures 6.3a, b show panoramic views during a normal flow and flooding, respectively, at St. 1.

For the flooding survey, flow observations and water quality analyses were conducted once or twice a year, mainly during the flooding season (May to October), for a total of 15 observations and analyses. One survey was conducted four times at the start of the survey, at the peak turbidity, at the peak flow rate and at the end of the survey, and a total of 60 samples were used for the water quality analysis. Water was sampled from the riverbank, placed into a plastic bottle using a ladle and taken back to the laboratory for analysis. The analyzed factors were the same as those in the normal flow survey. For the flow observations, a water level indicator was installed in advance, and the relationship between water depth and cross-sectional area was obtained by surveying the riverbed topography. At the time of the survey, the water depth was measured by a water level indicator, and the flow velocity was measured by a flow meter.

For St. 2 and St. 3, Fukui Prefecture measured the water quality of public water bodies from 1972 to 2015 (Fukui Prefecture Safety and Environment Department Environmental Policy Division 2021). The overall changes in total nitrogen, total phosphorus and chemical oxygen demand were summarized for the normal flow survey.

a

b

Fig. 6.3 Panoramic view during **a** a normal flow and **b** flooding at St. 1 (*Photos* by Akira Matsui). *Notes* **a** Photo date: May 20, 2016. View from downstream to upstream. **b** Photo date: September 20, 2016. View from downstream to upstream. *Source* Reprinted from Matsui (2018). Copyright 2018 Ecology and Civil Engineering Society

6.3 Results

6.3.1 Kochi River Flow and Water Quality

Table 6.1 and Fig. 6.4 show the median discharge and water quality during normal flows and flooding for 2006–2016 at St. 1. Both the discharge and water quality were significantly higher during flooding than during normal flows; the discharge was 15.4 times greater, the suspended solids were 217.5 times greater, the turbidity was 142.5 times greater, the chlorophyll a was 9.1 times greater, the total organic carbon was 6.1 times greater, the total nitrogen was 2.7 times greater, the total phosphorus was 16.5 times greater, and the chemical oxygen demand was 28.5 times greater. As a result of the Mann–Whitney U test, all p values confirmed significant differences.

6.3.2 Water Quality of the Kochi River, Kita River and Obama Bay

Figures 6.5, 6.6 and 6.7 show the overall changes in total nitrogen, total phosphorus and chemical oxygen demand during normal flows at St. 1, St. 2 and St. 3. Regarding the total nitrogen, St. 1 and St. 2 had almost the same value, but St. 3 had a significantly smaller value. Regarding total phosphorus, the trend was similar to that of total nitrogen, but St. 3 had values similar to those at St. 1 and St. 2. Regarding the chemical oxygen demand, St. 2 and St. 3 tended to have almost the same values, but St. 1 tended to have the lowest values.

6.4 Discussion

6.4.1 Relationship Between Discharge and Water Quality During Flooding and Normal Flows in the Kochi River

Table 6.2 and Fig. 6.8 show the relationship between discharge and water quality during flooding at St. 1. A strong correlation was confirmed between the discharge and suspended solids, turbidity, total phosphorus and chemical oxygen demand. A weak correlation was confirmed between the discharge and chlorophyll a and total nitrogen. A negative correlation was confirmed between the discharge and total organic carbon.

The cause of these results was that the inorganic matter topsoil was removed due to heavy rainfall, and it flowed out in proportion to the increase in discharge. On the other hand, organic matter (phytoplankton, etc.) was mainly deposited on

Table 6.1 Median discharge and water quality during normal flows and flooding for 2006–2016 at St. 1

	Discharge (m³/s)	Suspended solids (mg/L)	Turbidity (Nephelometric turbidity unit)	Chlorophyll a (μg/L)	Total organic carbon (mg/L)	Total nitrogen (mg/L)	Total phosphorus (mg/L)	Chemical oxygen demand (mg/L)
Normal flows	0.690	2	2	0.8	0.40	0.815	0.026	1.0
Flooding	10.655	435	285	7.3	2.45	2.200	0.430	28.5
P value	1.16e−16	3.37e−18	3.2e−18	5.42e−16	5.04e−18	1.13e−16	3.84e−18	3.73e−18

Note P value is the result of the Mann–Whitney U test
Source Modified from Matsui (2018). Copyright 2018 Ecology and Civil Engineering Society

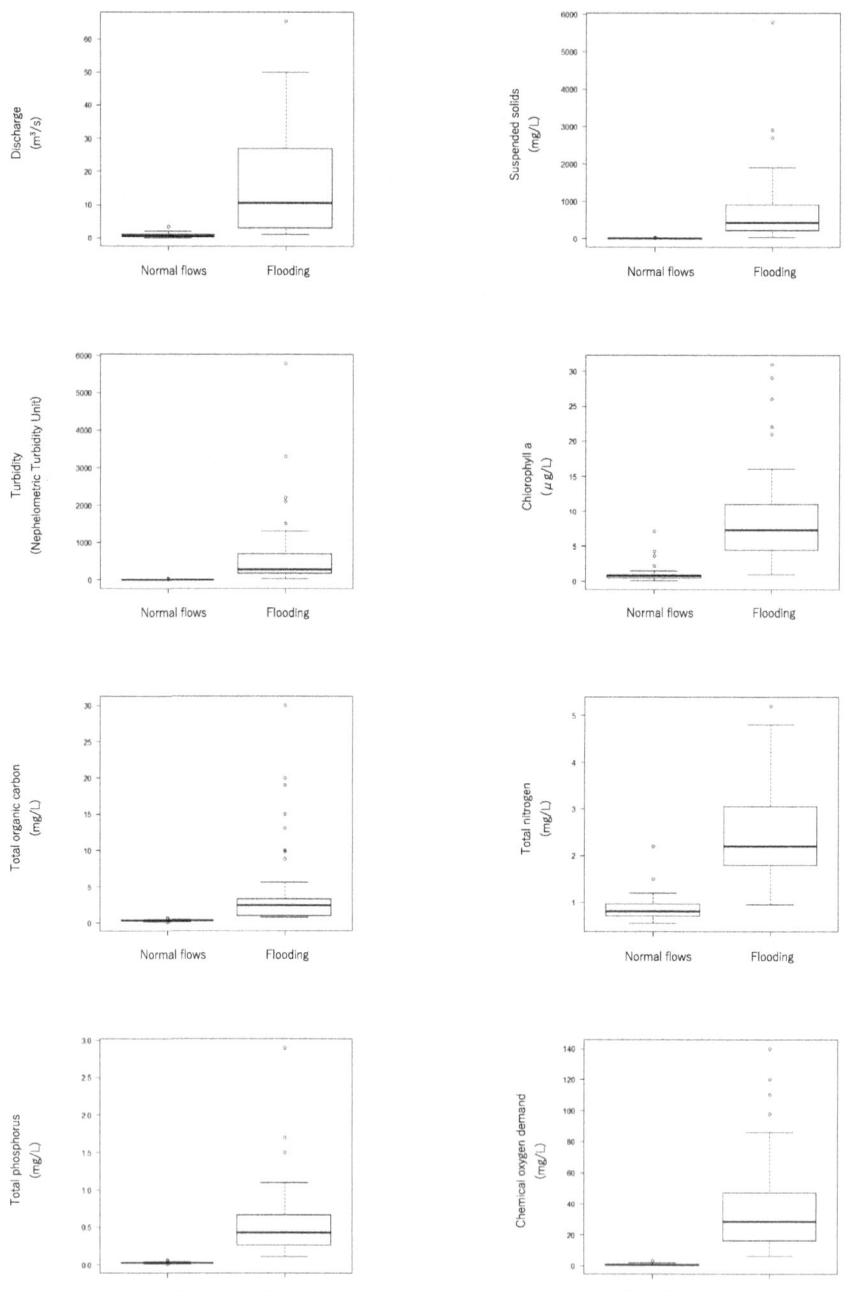

Fig. 6.4 Median discharge and water quality during normal flows and flooding for 2006–2016 at St. 1. *Source* Illustration based on Matsui (2018)

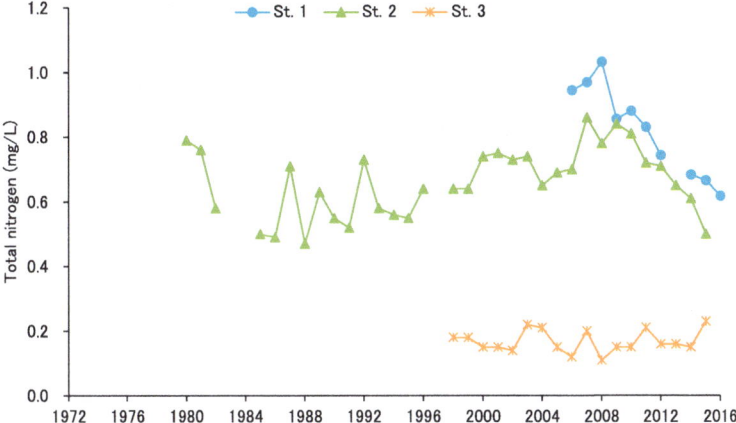

Fig. 6.5 Overall change in total nitrogen during normal flows at St. 1, St. 2 and St. 3. *Source* Reprinted from Matsui (2018). Copyright 2018 Ecology and Civil Engineering Society

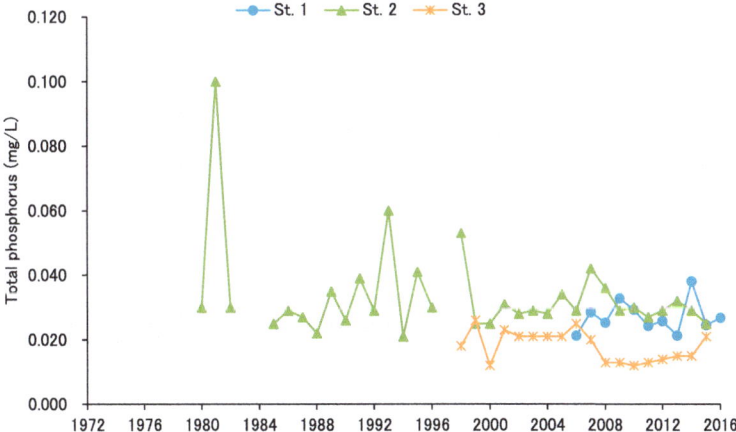

Fig. 6.6 Overall change in total phosphorus during normal flows at St. 1, St. 2 and St. 3. *Source* Reprinted from Matsui (2018). Copyright 2018 Ecology and Civil Engineering Society

the riverbed and flowed downstream. Then, the organic matter scarcely flowed out despite the increase in discharge.

Table 6.3 and Fig. 6.9 show the relationship between discharge and water quality during normal flows at St. 1. A weak correlation was confirmed between the discharge and suspended solids and turbidity. A negative correlation was confirmed between the discharge and total organic carbon. No correlation was confirmed for chlorophyll a, total nitrogen, total phosphorus and chemical oxygen demand.

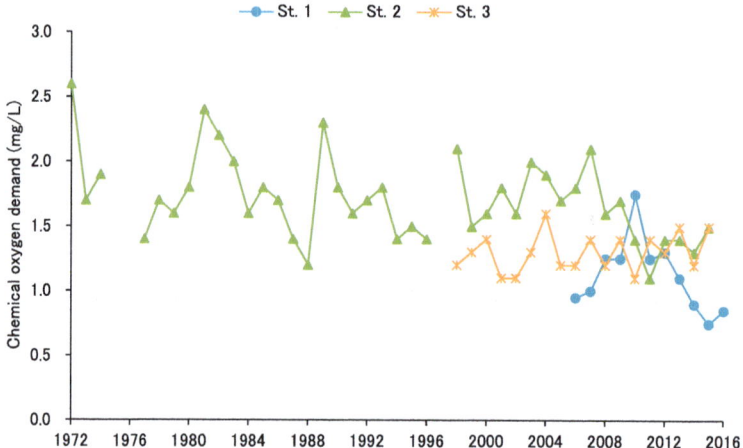

Fig. 6.7 Overall change in chemical oxygen demand during normal flows at St. 1, St. 2 and St. 3. *Source* Reprinted from Matsui (2018). Copyright 2018 Ecology and Civil Engineering Society

Table 6.2 Relationship between discharge and water quality during flooding for 2006–2016 at St. 1

	Suspended solids (mg/L)	Turbidity (Nephelometric turbidity unit)	Chlorophyll a (μg/L)	Total organic carbon (mg/L)	Total nitrogen (mg/L)	Total phosphorus (mg/L)	Chemical oxygen demand (mg/L)
Spearman's rank correlation coefficient	0.668	0.607	0.385	−0.248	0.400	0.664	0.427
P value	0.000	0.000	0.002	0.055	0.001	0.000	0.000

Note P value is the result of the Spearman's rank test
Source Modified from Matsui (2018). Copyright 2018 Ecology and Civil Engineering Society

It is presumed that the reason for these results was that the discharge was minimal during normal flows and the water quality was affected by seasonal changes rather than discharge fluctuations.

6.4.2 Relationship Among the Kochi River, Kita River and Obama Bay

Total nitrogen and total phosphorus are supplied from the forested areas and paddy fields in the upper and middle basins, and these nutrients are then returned to Obama Bay. On the other hand, the amount of chemical oxygen demand supplied from the upstream area is small, and it is thought that it is supplied from the middle and

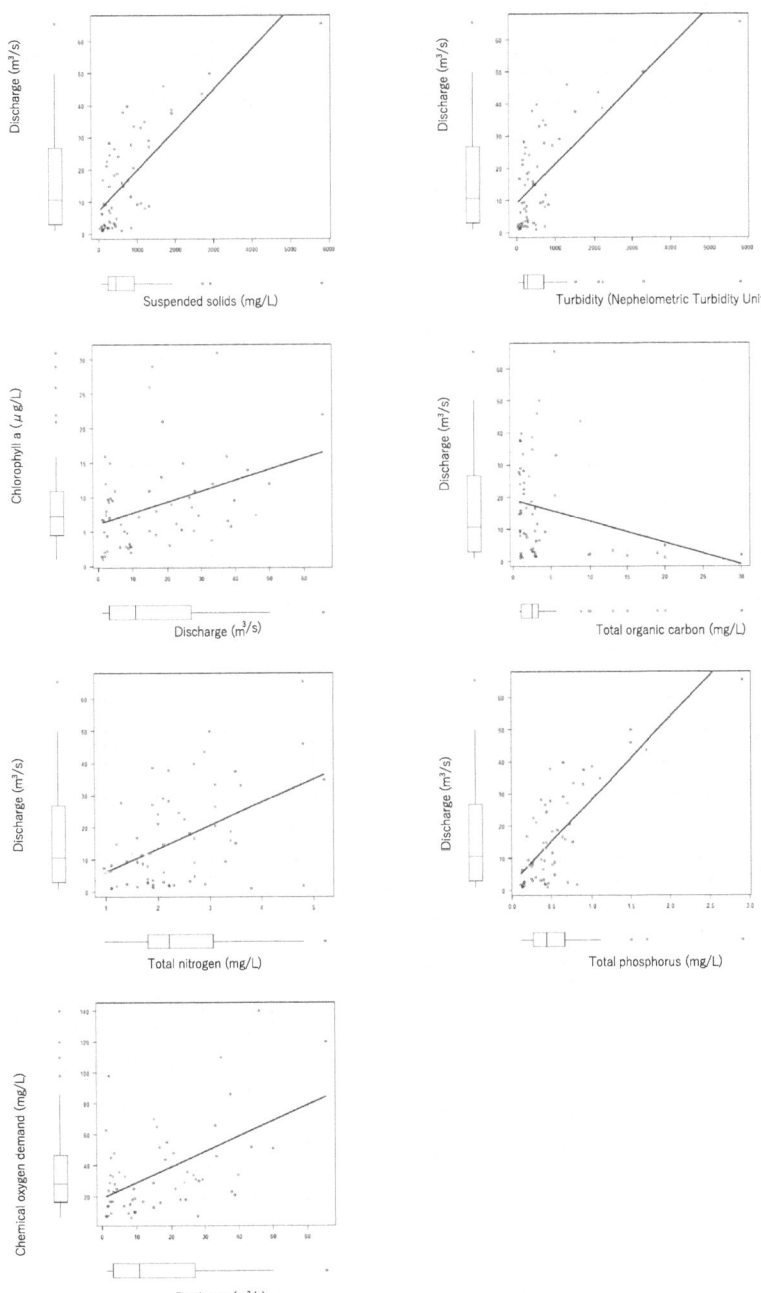

Fig. 6.8 Relationship between discharge and water quality during flooding for 2006–2016 at St. 1. *Source* Illustration based on Matsui (2018)

Table 6.3 Relationship between discharge and water quality during normal flows for 2006–2016 at St. 1

	Suspended solids (mg/L)	Turbidity (Nephelometric Turbidity Unit)	Chlorophyll a (μg/L)	Total organic carbon (mg/L)	Total nitrogen (mg/L)	Total phosphorus (mg/L)	Chemical oxygen demand (mg/L)
Spearman's rank correlation coefficient	0.358	0.458	0.131	−0.423	0.189	0.174	0.030
P value	0.016	0.001	0.402	0.004	0.219	0.258	0.846

Note P value is the result of the Spearman's rank test
Source Modified from Matsui (2018). Copyright 2018 Ecology and Civil Engineering Society

downstream areas. Yamada et al. (1998) reported that abundant nutrients flow out during normal flows in forested rivers in the catchment area of the Izarigawa Dam and Lake in the Ishikari River system in Hokkaido. Yamada and Inoue (2004) stated that the total phosphorus of the Umeda River, Hamada River and small farmland rivers flowing through Toyohashi City, Aichi Prefecture, remarkably exceeds the environmental standard values of Mikawa Bay, which is the downstream receiving area. In summary, nitrogen compounds are mainly supplied from forested areas, and phosphorus compounds are mainly supplied from paddy fields.

6.4.3 Impact of Floods on Closed Waters

In this study, during normal flows, organic matter (phytoplankton, nitrogen compounds and phosphorus compounds) flowed from the upper to middle to lower reaches and then into Obama Bay, a closed sea area. At St 1, during flooding in comparison with during normal flows, total nitrogen increased 2.7 times, total phosphorus increased 16.5 times and chemical oxygen demand increased 28.5 times.

Regarding inorganic matter (soil particles), suspended solids increased 217.5 times and turbidity 142.5 times greater during flooding than during normal flows, so a large amount of inorganic matter as well as organic matter flowed into Obama Bay.

From the above, the amount of organic and inorganic matter flowing downstream was much greater during flooding than during normal flows. Flooding endangers the lives and properties of humans living in basins but plays an important role in restoring and replenishing river ecosystems and supplying organic and inorganic matter to marine ecosystems.

In the reaches below the Arase Dam in the Kuma River Basin in Kumamoto Prefecture, total nitrogen was 16 times greater, total phosphorus was 74 times greater and chemical oxygen demand was 52 times greater during flooding than during normal flows (Hodoki et al. 2003). Compared with these results, those of this study showed that the rate of increase in these variables was small during flooding. The

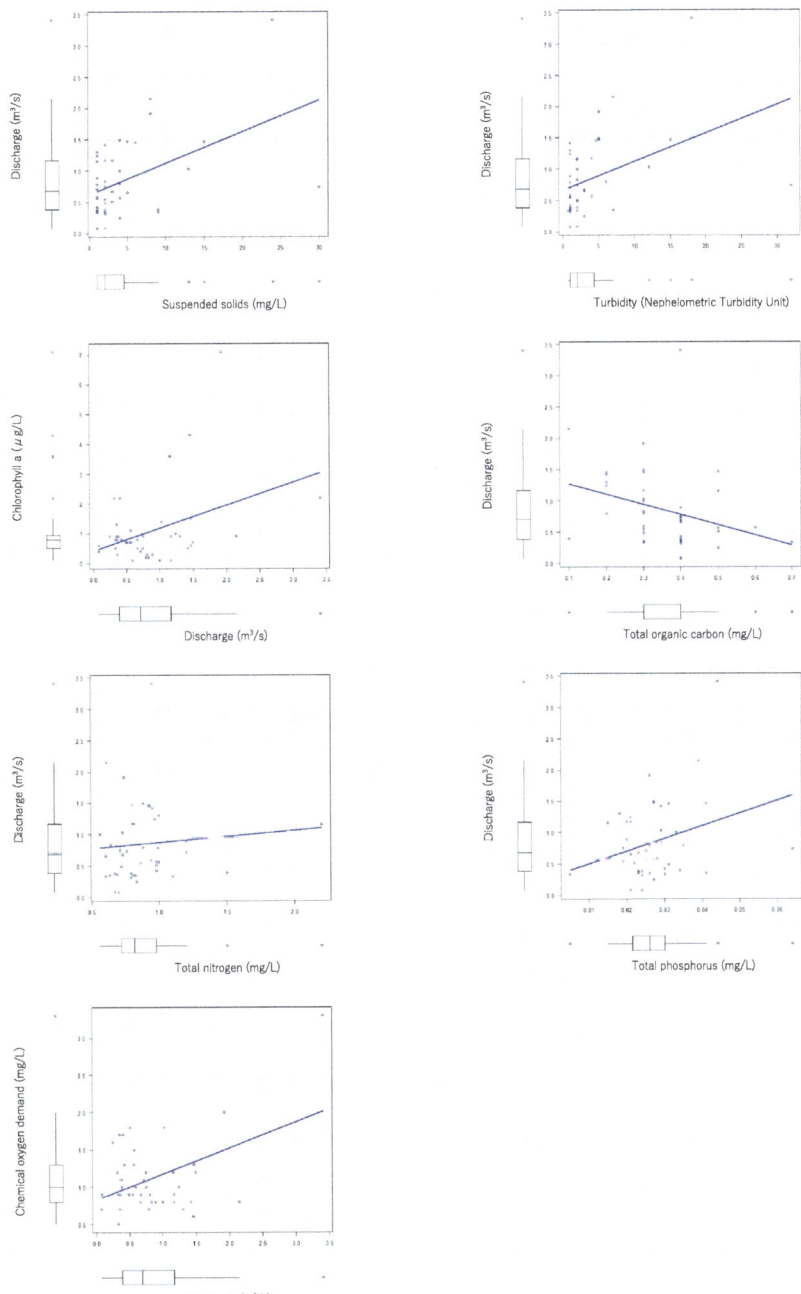

Fig. 6.9 Relationship between discharge and water quality during normal flows for 2006–2016 at St. 1. *Source* Illustration based on Matsui (2018)

reason for these results was likely that the abovementioned report was a survey after dam construction, whereas this research was a survey before the construction of a dam. The construction of the dam involves storying not only water but also earth, sand and nutrients. The nutrients are trapped because freshwater phytoplankton ingest nutrients and sink to the bottom of the dam. Thus, water with a lower nutrient concentration was supplied to the sea area located downstream after the construction of the dam (Yamamoto 2014). On the other hand, the sediment and nutrients that flowed into the dam flowed downstream without accumulating on the bottom of the dam during flooding, so it is possible that the concentration of nutrients was larger than that during normal flows.

Another possible cause for the difference in results is that sludge-like sediments near the bottom of the Arase Dam could have been picked up during flooding and discharged downstream with the river water. In addition, since there are multiple dams, such as the Setoishi Dam, Kawabegawa Dam and Ichifusa Dam, which are upstream of the Arase Dam in the Kuma River Basin, the influence of dam construction may have been significant.

To determine which of these multiple causes were most plausible, it was important to conduct the same survey after construction as before construction of the Kochigawa Dam and confirm the change in the rate of increase shown in Table 6.1.

The reasons that the turbid water flowing out of the paddy fields into Obama Bay was estimated as having a minimal effect on water quality and bottom sediment are the following: (1) The river inflow load factor was relatively small, (2) groundwater outflow contributed and (3) there was no large dam in the basin (Matsui 2014). Therefore, it is necessary to monitor what kind of changes in the water quality and bottom sediment of Obama Bay will occur due to the construction of the Kochigawa Dam.

In addition, decreases in fish catches in Yatsushiro Sea were more pronounced in areas that were affected by changes in the Kuma River. This scenario indicates that the decrease in fish catches in the sea area was likely to be closely related to some change in the Kuma River, that is the construction of the dam (Unoki 2004). Thus, the construction of the dam was related to changes in the load of nutrients reaching the sea and sedimentation on the dam. Similar to that in Yatsushiro Sea, the annual fish catches in Obama Bay were 1487 tons in 1993 but declined to 1000 tons in 2010 (Obama City 2014). However, as this phenomenon occurred before the construction of the Kochigawa Dam, it was not affected by the dam. To investigate the cause of the decrease in fish catches in Obama Bay, the same flow observations and water quality analyses are proposed, and the amount of sediment deposited on the dam after the construction of the dam should be monitored.

Box 6.1 Nature-Oriented River Management

Nature-oriented river management means preserving and creating habitats and diverse landscapes similar to those that naturally exist and managing the river to provide hydraulic and water utilization functions and environmental functions. For nature-oriented river management, it is important to consider the natural movements of the entire river in conjunction with local resident's daily life, history and culture.

For river improvements based on nature-oriented river management, maintenance of natural environmental functions that exist prior to the improvements should be considered as much as possible. Therefore, we are trying to use materials such as plants, trees and stones that exist at the construction site and to preserve the flow velocity, river width, water depth and vegetation changes in accordance with nature-oriented river management (Appendix Fig. 6.10).

Appendix

See Fig. 6.10.

a

b

Fig. 6.10 Example of nature-oriented river construction in the Itachi River in Yokohama City; **a** before construction and **b** after construction. *Source* Modified from the National Institute for Environmental Studies, https://tenbou.nies.go.jp/science/description/detail.php?id=95, Accessed October 31, 2021 (in Japanese)

References

Fukui Prefecture Safety and Environment Department Environmental Policy Division (2021) Water quality measurement results report for public water bodies and groundwater (in Japanese). http://www.erc.pref.fukui.jp/sogo/d094/. Accessed 22 June 2021

Hodoki Y, Sasaki K, Unoki S (2003) Water quality prediction and its problems at Kawabe River Dam (in Japanese). Rep Nat Conserv Soc Japan 94:31–46

Kochigawa Dam Construction Office, Reinan Promotion Bureau, Fukui Prefecture (2010) Kitagawa Comprehensive Development Project Kochigawa Dam Pamphlet (in Japanese)

Matsui A (2014) Effects of runoff water from paddy fields on enclosed coastal seas (in Japanese). Water Land Environ Eng 82:777–780. https://doi.org/10.11408/jjsidre.82.10_777

Matsui A (2018) Discharge and water quality during normal flow and flooding from 2006 to 2016 in Kochi River, Wakasa Town, Fukui Prefecture, Japan (in Japanese with English Abstract). Ecol Civ Eng 20:237–243. https://doi.org/10.3825/ece.20.237

Obama City (2014) About Obama City Sea Town Development Council (in Japanese). http://www1.city.obama.fukui.jp/obm/umimachi/HP/pdf.pdf. Accessed 24 June 2021

Sumi T (1997) Sand discharge from the dam reservoir and effluent quality management during sand discharge (in Japanese). Eng Dams 127:30–38

Unoki S (2004) Effects of the Kuma River Dams on the environment and fishery in Yatsushiro Bay (in Japanese with English Abstract). Oceanogr Japan 13:301–314. https://doi.org/10.5928/kaiyou.13.301

Yamada T, Inoue T (2004) Phosphorus outflow from farmland. Hokkaido Univ Collect Sch Acad Papers 12:57–60 (in Japanese)

Yamada T, Oe F, Shimizu T, Tachibana H (1998) A comparison of the runoff characteristics of nutrient loads from forest basins (in Japanese with English Abstract). Environ Eng Res 35:85–93. https://doi.org/10.11532/proes1992.35.85

Yamamoto T (2014) About oligotrophic conditions in the Seto Inland Sea (reconsideration) (in Japanese). J Jpn Inst Marin Eng 49:496–501. https://doi.org/10.5988/jime.49.496

Chapter 7
Open Levee Survey

Abstract The Kita River basin has 11 open levees. An open levee returns flood waters to the river to prevent flood damage from spreading. Most of the areas surrounding the open levee are used as paddy fields, although some areas have been developed in recent years, and the paddy fields have converted into residential land. As a consequence, houses in the residential area are being damaged by floods. In the area of the open levee, the agricultural working population is declining and aging. It is possible that the damage caused by large-scale flood disasters in recent years has expanded due to the loss of residents' dependence in the basin.

Keywords Aging · Dependence · Flood disaster · Open levee · Paddy field · Resident

7.1 Introduction

Open levees exist many in rural areas and are discontinuous embankments in which an opening exists on the downstream side. An open levee returns floodwater to a river to prevent the spread of damage. Open levees can be classified into three levee types according to their roles because they are similar in morphology but completely different in function depending on whether they are in a fast-flowing river or a slow-flowing river (Fig. 2.2).

The number of open levees increased until the early Showa period and decreased after the war. Since projects under direct control in Japan are generally planned on the premise that floods will not flow over levees, open levees were not necessary and were removed. As a result, discontinuous embankments such as an open levee are now being renovated and transformed into a continuous embankment (Teramura and Okuma 2005).

However, among the first-class rivers, there are river basins where constant floods occur due to the remaining open levees. In such rivers, the residents hope that the open levees are immediately closed and a continuous embankment is constructed to improve the land use inside the embankment (Haruyama 2001).

© The Author(s), under exclusive license to Springer Nature Singapore Pte Ltd. 2022
A. Matsui, *Wetland Development in Paddy Fields and Disaster Management*,
https://doi.org/10.1007/978-981-19-3735-4_7

On the other hand, due to the influence of the River Act Amendment in 1997, technologies based on the uniform standards and traditional techniques with regional characteristics, for example open levees, are being reviewed.

In recent years, record heavy rainfall has increased due to the effects of climate change and caused large-scale flooding of rivers. The national government has proposed watershed control measures, such as improving storage facilities and strengthening evacuation systems, to improve flood safety throughout the entire basin. The state has noted that floods cannot be prevented by dams and embankments alone.

Therefore, before heavy rains, the water used for agriculture and power generation must be flushed to increase the vacant areas of dams. The capacity to store rainwater will be increased by developing storage facilities in the basin and by utilizing paddy fields and ponds. Additionally, open levees will be constructed so that water can flow safely even if the water level increases.

7.2 Methods

This section discusses the Kita River basin, which has 11 open levees (Fig. 7.1). The Kita River originates near Sanjusangenzan in the Nosaka Mountains, which borders

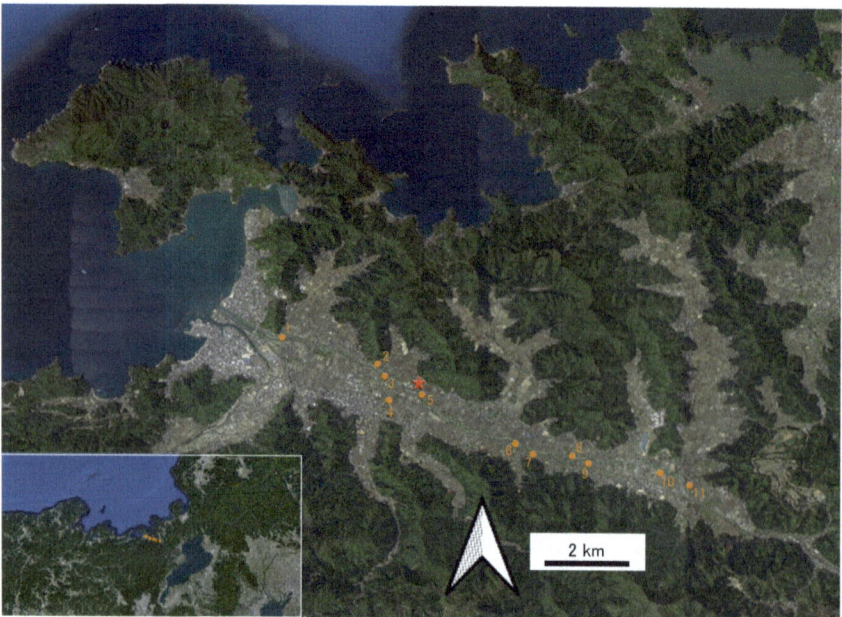

Fig. 7.1 Location of open levees. *Note* The background is a Google Earth image

the Shiga and Fukui prefectures, and flows southward through the mountains of Takashima City, Shiga Prefecture, and in Wakasa Town, Fukui Prefecture.

It is a first-class river with a trunk river channel length of 30.3 km and a basin area of 210.2 km^2 that flows into Obama Bay, Sea of Japan. The relatively large inflow rivers are the Eko River, Onyu River, Matsunaga River, Nogi River, Sugiyama River and Toba River. Land use in the basin is approximately 83% forests, approximately 13% agricultural land such as paddy fields and upland fields and approximately 4% other land use classes such as residential lands.

This basin is characterized by a rapid increase in water when heavy rains fall and a decrease in water level as soon as heavy rains stop because the Kita River has a small basin area, a short channel length and a steep slope. Since it is financially difficult to build a continuous and strong embankment, an open levee system has been adopted to prevent temporary floods from being stored in paddy fields and to protect the urban area of Obama City.

The current status of 11 open levees in the Kita River basin was investigate. The eleven areas with open levees were classified according to Fig. 2.2. Problems with open levees were identified, and solutions and future strategies were examined.

7.3 Results

Of the 11 open levees, 1–5 are located in Obama City, Fukui Prefecture, and 6–11 are located in Wakasa Town, Fukui Prefecture. The gradients at all 11 locations were steeper than 1/1,000. All 11 locations were classified as the drainage levee type (Fig. 2.2). The paddy fields adjacent to the open levees were already developed, and there was little abandoned cultivated land.

Figures 7.2, 7.3, 7.4 and 7.5 show 11 fixed-point photographs of open levees. They are photographs of the four seasons with normal flows. The green that is around the open levees indicates that the paddy fields have been filled with water and rice cultivation has begun (Fig. 7.4). Similarly, the golden color that is around the open levees indicates that the rice grew and it is before or after harvesting (Fig. 7.5).

7.4 Discussion

7.4.1 Intensification of Flood Disasters

Due to Typhoon No. 18 in September 2013, one embankment in the Nogi River broke down (Figs. 7.1 ★ and 7.6). Precipitation at the time of the washout exceeded 200 mm/day (Fig. 7.7). Typhoon No. 21 in October 2017 also exceeded 200 mm/day, but at this time, the embankment did not washout.

Fig. 7.2 Panoramic view of 11 open levees in autumn (*Photos* by Akira Matsui). *Note* Photo date: October 14, 2020

Fig. 7.3 Panoramic view of 11 open levees in winter (*Photos* by Akira Matsui). *Note* Photo date: February 26, 2021

Fig. 7.4 Panoramic view of 11 open levees in spring (*Photos* by Akira Matsui). *Note* Photo date: May 26, 2021

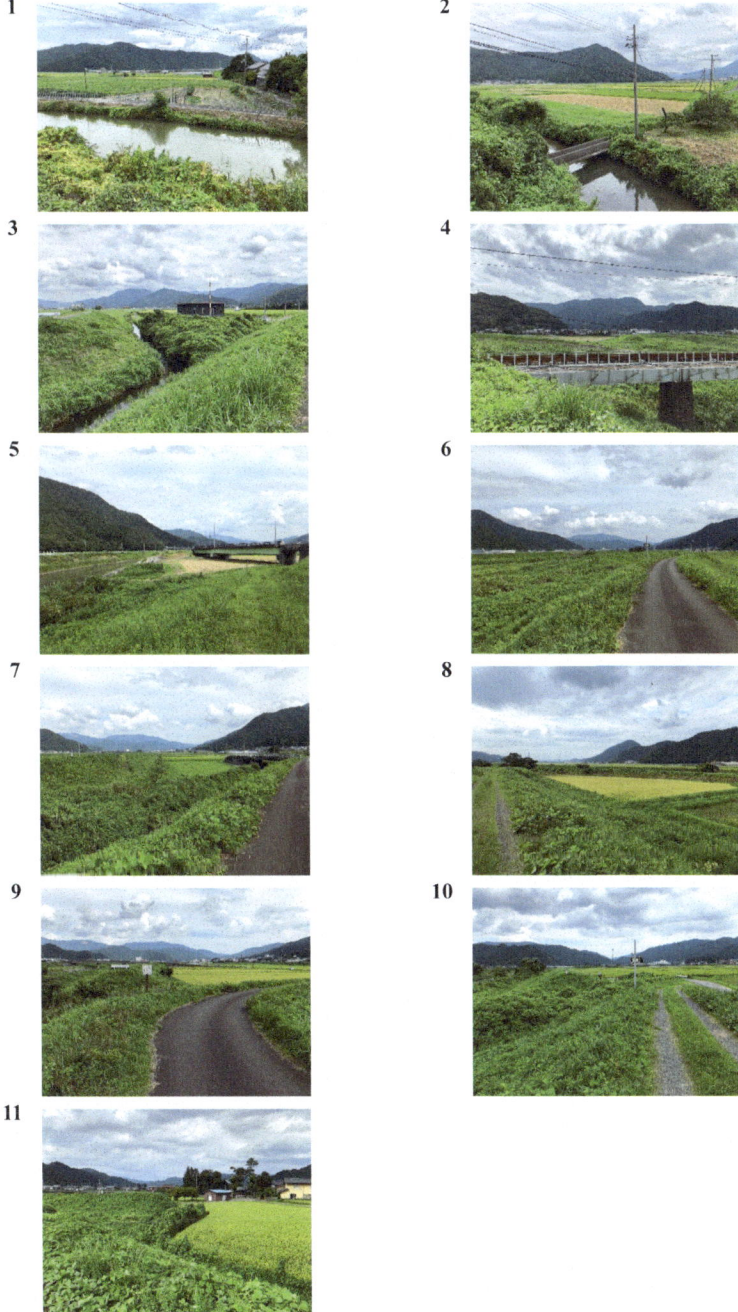

Fig. 7.5 Panoramic view of 11 open levees in summer (*Photos* by Akira Matsui). *Note* Photo date: August 26, 2021

Fig. 7.6 Washout in the Nogi River. *Source* Reprinted from the Kinki Regional Development Bureau, Ministry of Land, Infrastructure, Transport and Tourism of Japan, https://www.kkr.mlit.go. jp/plan/saigairaiburari/2013_t18/2013_t18_0043.html, Accessed October 31, 2021 (in Japanese)

During the 10 years from 2010 to 2020, precipitation exceeded 100 mm/day 17 times and exceeded 200 mm/day twice. If a large amount of precipitation is recorded in a short time due to the influence of abnormal weather in the future years, then even if there are 11 open levees in the Kita River basin, the risk of a flood-related disaster is unavoidable.

The maximum daily precipitation at the time of the washout was 253.5 mm. Of the 20 years from 2001 to 2020, daily precipitation of 100 mm or more was recorded on 3 days from 2001 to 2010 and 15 days from 2011 to 2020 in Obama City, Fukui Prefecture, Japan (Fig. 7.8). After testing whether there was a difference in the median between these two groups, a significant difference was confirmed (Wilcoxon's rank sum test, $p = 0.011$). Thus, the flowing water in the Kita River was swiftly discharged downstream during past precipitation events. However, when abnormal precipitation such that the daily precipitation exceeded 100 mm in recent years, the water in the Kita River stopped flowing and began to flow back into the Nogi River. As a result, the water in the Nogi River stopped flowing, and it is probable that the embankment near open levee 5 broke.

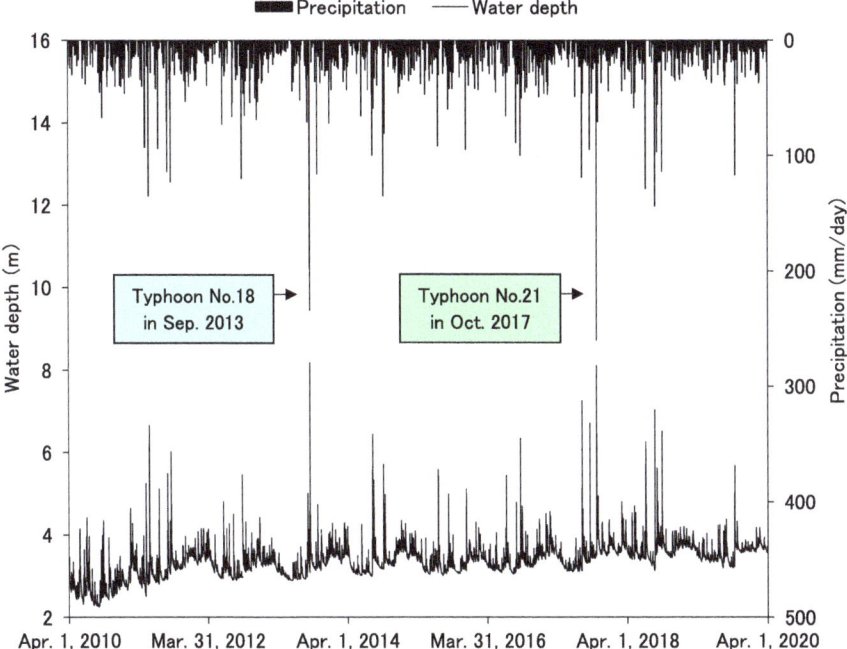

Fig. 7.7 Overall change in precipitation and water depth in the Kita River. *Source* Illustration based on water level: Takatsuka Observatory, Ministry of Land, Infrastructure, Transport and Tourism, Precipitation: Obama Observatory, Japan Meteorological Agency, https://www.jma.go.jp/jma/menu/menureport.html, Accessed October 31, 2021 (in Japanese)

7.4.2 Agricultural Land Conversion

Most of the areas surrounding the open levees are used as agricultural lands. However, in recent years, the area surrounding open levee 1 in the Eko River has been developed, and farmland has been converted into residential land. Thus, houses have experienced flood damage (Fig. 7.9). This tendency was confirmed in not only the downstream area but also the middle and upstream areas, such as at open levee 11 (Fig. 7.10).

7.4.3 Declining and Aging of Agricultural Working Population

The paddy field dams (Fig. 2.4) are effective at mitigating flood risks. In addition, the development of irrigation and drainage canal systems will prevent the destruction of cultivated land due to flooding and seepage. These mitigating effects are produced by farming the paddy fields. However, in recent years, it has become difficult to manage

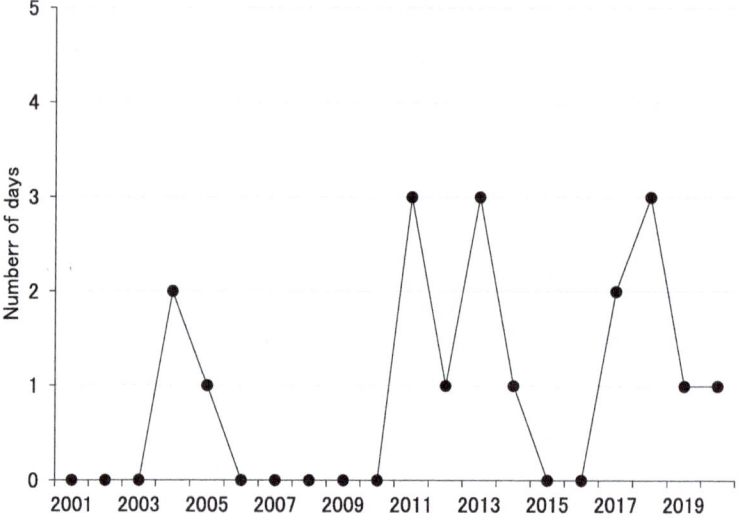

Fig. 7.8 Number of days with daily precipitation of 100 mm or more in Obama City, Fukui Prefecture, Japan. *Source* Modified from the Japan Meteorological Agency, https://www.data.jma.go.jp/obd/stats/etrn/index.php, Accessed December 8, 2021 (in Japanese)

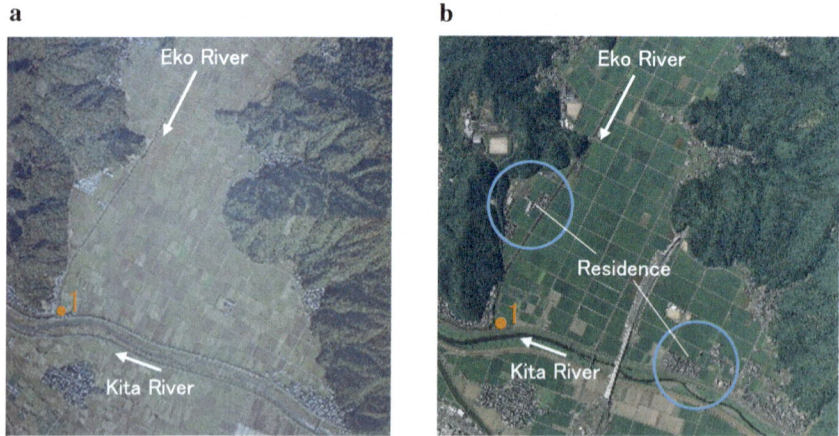

Fig. 7.9 Land use changes in the area surrounding open levee 1. *Note* **a** 1975 and **b** 2013. *Source* Reprinted from the Geospatial Information Authority of Japan, https://mapps.gsi.go.jp/maplibSearch.do#1, Accessed October 31, 2021 (in Japanese)

paddy field agriculture due to the declining and aging population of agricultural workers.

The agricultural working population showed a declining trend in 2015, and the proportion of those over 65 years old accounted for 78% and 77% in Obama City and Wakasa Town, respectively (Fig. 7.11). As a result, the rate of decrease in cultivated

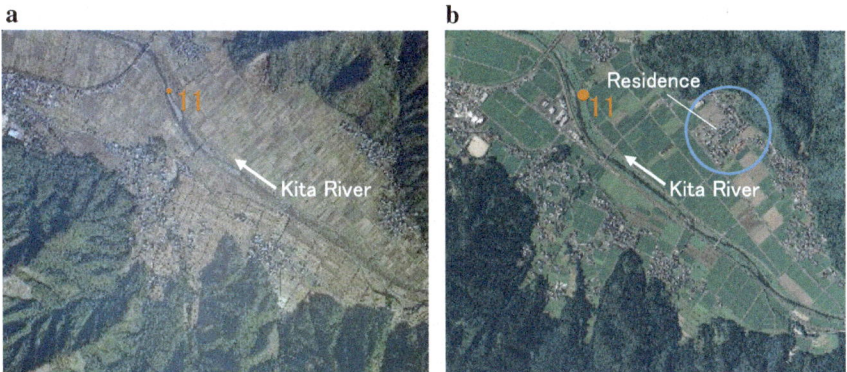

Fig. 7.10 Land use changes in the area surrounding open levee 11. *Note* **a** 1975 and **b** 2013. *Source* Reprinted from the Geospatial Information Authority of Japan, https://smapps.gsi.go.jp/maplibSea rch.do#1, Accessed October 31, 2021 (in Japanese)

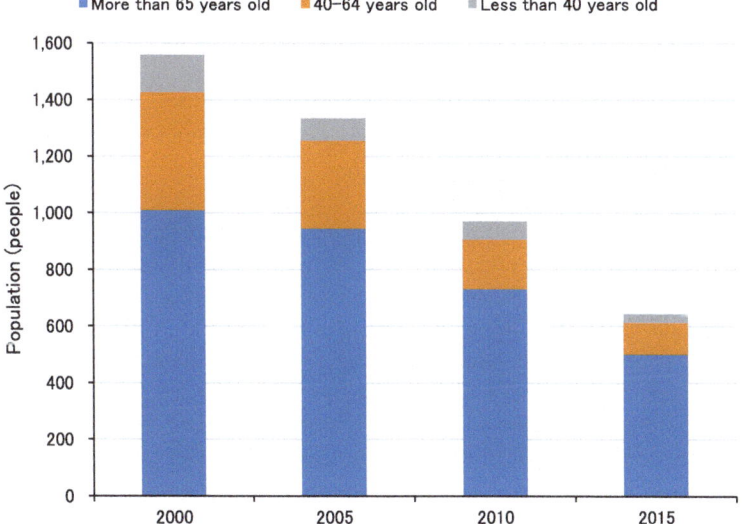

Fig. 7.11 Overall change in age-specific agricultural working population in Obama City, Fukui Prefecture, Japan. *Source* Illustration based on the Ministry of Agriculture, Forestry and Fisheries of Japan (2020)

land area in 2015 compared to that in 2000 was 19% in Obama City and 10% in Wakasa Town, respectively (Fig. 7.12). It is estimated that the construction of residential land centered around the city area has had an impact in Obama City.

From the above information, it is likely that the cause of the washout of the open levee in the Nogi River was that the amount of rainwater permeating underground

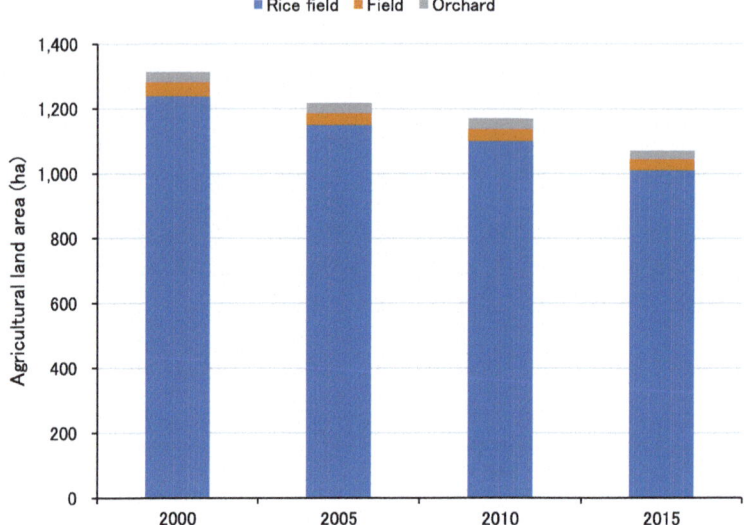

Fig. 7.12 Overall change in cultivated area in Obama City, Fukui Prefecture, Japan. *Source* Illustration based on the Ministry of Agriculture, Forestry and Fisheries of Japan (2020)

decreased due to the decrease in cultivated land area and the increase in residential land in the entire Kita River basin. Thus, short-duration rainfall events have remarkably increased in recent years.

To address these problems, it is necessary to take measures prevent the development of residential land in areas near open levees where inundation damage is expected. The Ministry of Finance has proposed that disaster risk reduction costs should be reduced by limiting the development of land at high risk of damage and encouraging people to live in other areas identified by the Financial System Council. The Ministry of Land, Infrastructure, Transport and Tourism has provided guidelines for disseminating information on local flood hazards. Obama City is trying to accumulate farmland by increasing the number of new farmers that can start farming and establishing village farming organizations such as limited companies, agricultural cooperative corporations and general incorporated associations.

7.4.4 Restoration of Basin Residents' Dependence on the River

The modernization of river technology has made it easier for embankments to be constructed. If a solid embankment has been developed, then the area inland of the embankment is less likely to overflow even during floods. As a result, it is no longer necessary for basin residents to cooperate during a flood event or take measures to minimize flood damage. It is presumed that the residents of a basin are less involved with the river, and the dependence of residents on the river tends to disappear.

It is possible that the damage caused by large-scale flood disasters in recent years has expanded due to the loss of dependence of residents on the basin. By reducing the opportunities to be close to a river, the threat of a river cannot be felt. Residents do not notice the rise in water level because they have little interest in the rivers around them. As a result, the evacuation of residents in the basin is delayed.

The dependence of the basin residents on the river can be restored by conserving and further developing the open levee through the watershed control project currently advocated by the Ministry of Land, Infrastructure, Transport and Tourism. The conventional hydraulic control method will be changed from curbing floods to addressing overflowing water. For this type of watershed control to be successful, the former dependence of basin residents on the river must be restored.

7.4.5 Challenges to and Solutions for Open Levee Conservation

Changing a paddy field area to an urban area by changing a discontinuous embankment to a continuous embankment is not preferable from the viewpoint of basin management because it reduces the buffer zone during floods. It is important to leave paddy fields as retarding basins in terms of floodplain management (Haruyama 2001).

In the future, in cooperation with the relevant government agencies, it will be necessary to uniformly provide compensation to those in the retarding basin as a more substantial consideration for environmental conservation (Sugiura 2007). Residents in the middle and upper reaches (rural areas) who live near the open levee may be inundated with each heavy rain, and they are constantly struggling and feeling unfairly treated. Thus, the country should provide sufficient public compensation. Residents of the lower reaches (urban areas) must also be grateful to the residents of the middle and upper reaches.

The maintenance of open levees is effective for more than hydraulic control measures. The forgotten function of open levees is not only the reduction of flood-water but also the puddle that form during floods and serve as refuges for many fish and other organisms. During floods, the main stem of a river becomes a torrent, so without refuge sites, many fish and other organisms would be washed away to the sea (Okuma 2004). Therefore, the maintenance of open levees is also important as an environmental conservation measure.

As mentioned above information, although there are some difficulties in maintaining open levees, it is possible to protect lives from large-scale to small-scale flood disasters. By addressing flooding disasters throughout the basin, the dependence of the residents in the basin on the river can be restored, and at the same time, the use of water, such as agricultural water and regional water, and the benefits of the environment, such as fish conservation and recreation, will benefit. Thus, the benefits of open levees will greatly exceed the negative impacts.

In today's world where independence is respected, the demand for community will be resisted. However, to build a resilient community that is resistant to intensifying flood disasters, appropriate independent and community-based efforts are required.

Box 7.1 Nature Restoration Project

Approximately 180 km^2 of Kushiro Mire extends to the lower reaches of Kushiro River in Hokkaido, Japan. Kushiro Mire, the largest in Japan, is a habitat for valuable animals and plants. The full-scale hydraulic control project in the Kushiro River was triggered by the flood in 1920, and river improvements such as straightening the meandering river began. Such river improvements resulted in an increase in sediment inflow to Kushiro Mire and a decrease in the frequency of flooding. In addition, the area of the Kushiro Mire decreased by approximately 20% over the half century from 1947 to 2004, and its ecosystem was greatly affected, such as through the deterioration of the landscape and the decrease in rare wildlife.

With the enactment of Nature Restoration Promotion Law, the nature restoration project in Kushiro Mire was fully implemented in 2003. Through this nature restoration project, prevention of sediment inflow into the Kushiro Mire, restoration of meandering rivers (Appendix Fig. 7.13), restoration of the Kushiro Mire and planting trees in degraded lands are being implemented.

Appendix

See Fig. 7.13.

a

b

Fig. 7.13 Kushiro River in Hokkaido, Japan. **a** Channelized river and **b** remeandering river. *Note* **a** 2001 and **b** 2010. *Source* Reprinted from the Japan RiverFront Research Center. https://www.chr ome-extension://efaidnbmnnnibpcajpcglclefindmkaj/viewer.html?pdfurl=http%3A%2F%2Fwww. a-rr.net%2Fjp%2Fwaterside%2Fdomestic%2Fdocs%2F2011J01_kushiro.pdf&clen=902492& chunk=true. Accessed October 31, 2021 (in Japanese)

References

Haruyama S (2001) Land use planning deliberated on prevention against flooding (in Japanese). J Rural Plann 20:87–90. https://doi.org/10.2750/arp.20.87

Ministry of Agriculture, Forestry and Fisheries of Japan (2020) Report on results of Longitudinal Census of agriculture in Japan (in Japanese). https://www.maff.go.jp/j/tokei/census/afc/2020/. Accessed 30 Oct 2021

Okuma T (2004) Creating local ideas 1 Technology also has autonomy—Tradition and modernity of water control technology- (in Japanese). Rural Culture Association Japan, Tokyo

Sugiura M (2007) Co-existence with water in upper basin inhabitants along open levees Case study of farmers in Miyazaki Prefecture simultaneously pursuing environmental servation and flood control (in Japanese with English Abstract). J Jap Soc Hydrol Water Res 20:34–46. https://doi.org/10.3178/jjshwr.20.34

Teramura J, Okuma T (2005) A study on evolution and role of open levees on alluvial-fan rivers in the Hokuriku District -from the view point of decentralization of river-engineering decision making- (in Japanese with English Abstract). J Histor Stud Civ Eng 24:161–171. https://doi.org/10.11532/journalhs2004.24.161

Chapter 8
Survey of the River Ecosystem Downstream of a Dam

Abstract Dam construction is very effective in hydraulic control; however, dam construction also has negative impacts on river ecosystems. On the other hand, we do not understand how and why a dam influences riverine macroinvertebrate assemblages. The purpose of this study was to clarify the influences of the Ooishi Dam (Sekikawa Village, Niigata Prefecture, Japan) on downstream macroinvertebrate assemblages with a particular focus on net-spinning caddisflies. I conducted a monthly field survey at one site upstream and three sites downstream of the dam in the Ooichi River. The water quality, stone surface deposits and macroinvertebrate assemblage surveys were conducted for approximately 20 days every month from April–December 1994. The upstream macroinvertebrate assemblage was dominated by Ephemeroptera and Plecoptera, whereas the downstream assemblage was dominated by Trichoptera, especially net-spinning caddisflies in the genus *Stenopsyche*; *Stenopsyche marmorata* and *Stenopsyche sauteri*. Net-spinning caddisfly species were more abundant at the downstream sites than at the upstream site. First instar larvae of *Stenopsyche* sp. showed remarkably high density in summer at the downstream sites. The increase in sestonic organic matter concentration in the river water at the downstream sites in summer was presumed to be due to the flow down of the phytoplankton grown in the Ooishi Dam reservoir. The positive effect of the phytoplankton from the reservoir on the juvenile larvae of *Stenopsyche* sp. likely resulted in the increase in the number of individuals of net-spinning caddisflies at the downstream sites. Phytoplankton that originated from the reservoir might have enhanced the downstream population of *Stenopsyche* sp. This river ecosystem where the number of individuals of *Stenopsyche* sp. is remarkable is not normal. A true river ecosystem consists of various species that live equally. When a dam reservoir is constructed in the upstream area of a river, a dam control plan that preserves the natural ecosystem as much as possible must be applied.

Keywords Aquatic insect · Phytoplankton · River ecosystem · *Stenopsyche marmorate* · *Stenopsyche sauteri*

This chapter is a revised version of Matsui (2008). Copyright 2008 Ecology and Civil Engineering Society. https://doi.org/10.3825/ece.11.175. Accessed November 27, 2021.

© The Author(s), under exclusive license to Springer Nature Singapore Pte Ltd. 2022   95
A. Matsui, *Wetland Development in Paddy Fields and Disaster Management*,
https://doi.org/10.1007/978-981-19-3735-4_8

8.1 Introduction

Constructing dams disrupts biological and physical cycles and diminishes flow fluctuations with catastrophic damage to aquatic animals (Morishita 2001; Okuma 1999).
The most common changes in macroinvertebrate assemblages downstream of a dam
are increases in population density and biomass despite decreasing species diversity (Tanida and Takemon 1999). The factors that results in such change are (1) a
controlled flow rate; (2) a change in channel shape; (3) a change in water temperature
environment; (4) an occurrence of turbidity; (5) an increase in plankton that have
been produced in the dam reservoir; and (6) obstacles to the movement of macroinvertebrate assemblages. Thus, a number of dam-related factors affect downstream
macroinvertebrate assemblages. The cause and effect should be clarified; however,
there is no related case study in Japan.

   The construction of the Sameura Dam and the resulting rich supply of planktonic
algae probably enhanced the colonization and settlement of *Macrostemum radiatum*
(Furuya 1998). In this study, I focus on the influence of plankton that has been
produced in a dam reservoir. In addition, I compare the macroinvertebrate biomass
upstream and downstream of the Ooishi Dam (Sekikawa Village, Niigata Prefecture, Japan) and consider the effects of the dam on downstream macroinvertebrate
assemblages with a particular focus on net-spinning caddisflies. The purpose of this
study was to clarify the influence of dam construction on the river ecosystem downstream of the Ooishi Dam (38° 01′ 50″ N, 139° 34′ 05″ E) (Fig. 8.1), which involves a
surface water withdrawal system. Particular emphasis was given to changes in stream
regimes and water quality.

8.2 Methods

8.2.1 Survey Area

The Ooishi River is a tributary of the Ara River that flows into the Sea of Japan. The
Ooishi Dam was constructed for power generation purposes and is located approximately 7.5 km upstream from the confluence of the Ara River (Fig. 8.1). The Ooishi
Dam has the following characteristics: the dam height is 87.0 m; the crest height is
187.0 m; the flood full water level is 184.5 m; the constant full water level is 184.0 m;
the flood-limited water level is 155.0 m; the lowest water level is 154.0 m; and the
total water storage capacity is 22,800,000 m^3. Construction on this dam began in
1970 and was completed in 1978; it is a multipurpose dam that provides both flood
control and power generation. The specifications of the Ooishi Dam are presented in
Table 8.1.

   At Ooishi Dam, the maximum amount of water consistently flowing into a water
pressure iron pipe for electric generation (145.0 m in elevation) is 15 m^3/s. Electric
generation is provided using the surface water of the reservoir obtained through a

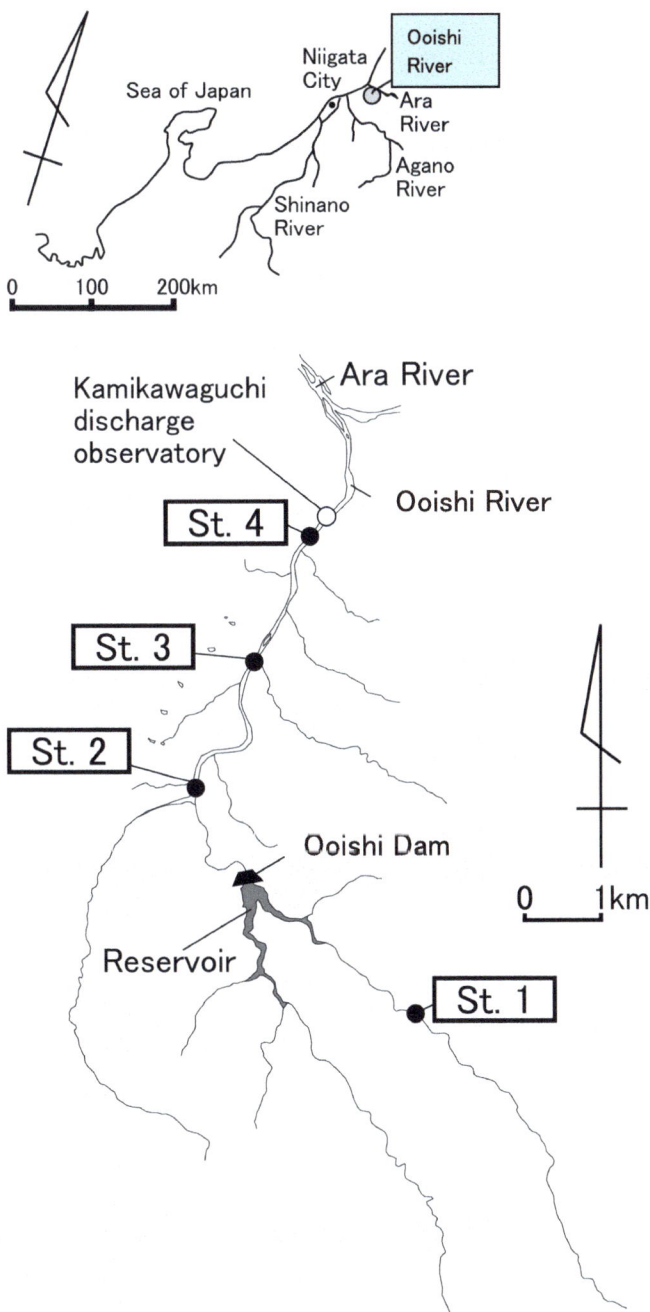

Fig. 8.1 Location of the survey sites. *Source* Reprinted from Matsui (2008). Copyright 2008 Ecology and Civil Engineering Society

Table 8.1 Specifications of the Ooishi Dam

Total reservoir storage (10^3 m^3)	Basin area (km^2)	Purpose	Dam height (m)	Water surface area (ha)	Completion of construction (year)
22,800	69.8	Flood control Power generation	87.0	110	1978

Source Reprinted from the Japan Dam Foundation, http://damnet.or.jp/cgi-bin/binranA/All.cgi? db4=0775, Accessed November 1, 2021 (in Japanese)

surface water intake gate. A conduit gate is used when the amount of discharge is more than 15 m^3/s, and crest gates are used in case of an emergency.

St. 1 is located 1.7 km upstream from the dam backwater (Fig. 8.2); St. 2 is located 1.7 km downstream of the dam site where the reservoir water flows directly through the water pressure iron pipe for electric generation (Fig. 8.3); St. 3 is positioned 1.7 km downstream of St. 2 near Kajikadani Bridge (Fig. 8.4); and St. 4 lies 2.1 km downstream of St. 3 near Kuratajima Bridge close to the confluence of the Ara River (Fig. 8.5). Table 8.2 shows the altitude, channel width, dominant bed material and river morphology (Kani 1944) of these study sites.

The seasonal changes in the hydrologic environments in the Ooishi River that occurred during this study are shown in Fig. 8.6. The annual amount of rainfall in

Fig. 8.2 Panoramic view at St. 1 (*Photo* by Akira Matsui). *Notes* Photo date: July 1994. View from downstream to upstream

Fig. 8.3   Panoramic view at St. 2 (*Photo* by Akira Matsui). *Notes* Photo date: July 1994. View from downstream to upstream

Fig. 8.4   Panoramic view at St. 3 (*Photo* by Akira Matsui). *Notes* Photo date: July 1994. View from downstream to upstream

Fig. 8.5 Panoramic view at St. 4 (*Photo* by Akira Matsui). *Notes* Photo date: July 1994. View from downstream to upstream

Table 8.2 Physical features of the survey sites

Survey sites	Altitude (m)	Channel width (m)	Dominated bed material	River morphology
St. 1	200	10	Cobbles	Aa
St. 2	100	30	Gravel and sand	Bb
St. 3	80	40	Gravel and sand	Bb
St. 4	60	50	Gravel and sand	Bb

Note River morphology was determined according to Kani (1944)
Source Reprinted from Matsui (2008). Copyright 2008 Ecology and Civil Engineering Society

1994 was 2017 mm and 78% of the mean in the past for 25 years. Both inflow discharge into the Ooishi Dam reservoir and outflow discharge into the Ooishi River increased due to the influence of snowmelt in early April and the rainy season in early July. A significant difference in the flow rate variation pattern between areas upstream and downstream of the dam was not observed. It is likely that the effects of the flow rate control of the dam were lessened by the inflow from tributaries.

Fig. 8.6 Seasonal changes
in the hydrologic
environments in the Ooishi
River during this survey.
Note **a** Daily precipitation at
Shimoseki, **b** inflow
discharge into the Ooishi
Dam reservoir, **c** outflow
discharge into the Ooishi
River and **d** discharge at the
Kamikawaguchi discharge
observatory (see Fig. 8.1).
Source Reprinted from
Matsui (2008). Copyright
2008 Ecology and Civil
Engineering Society

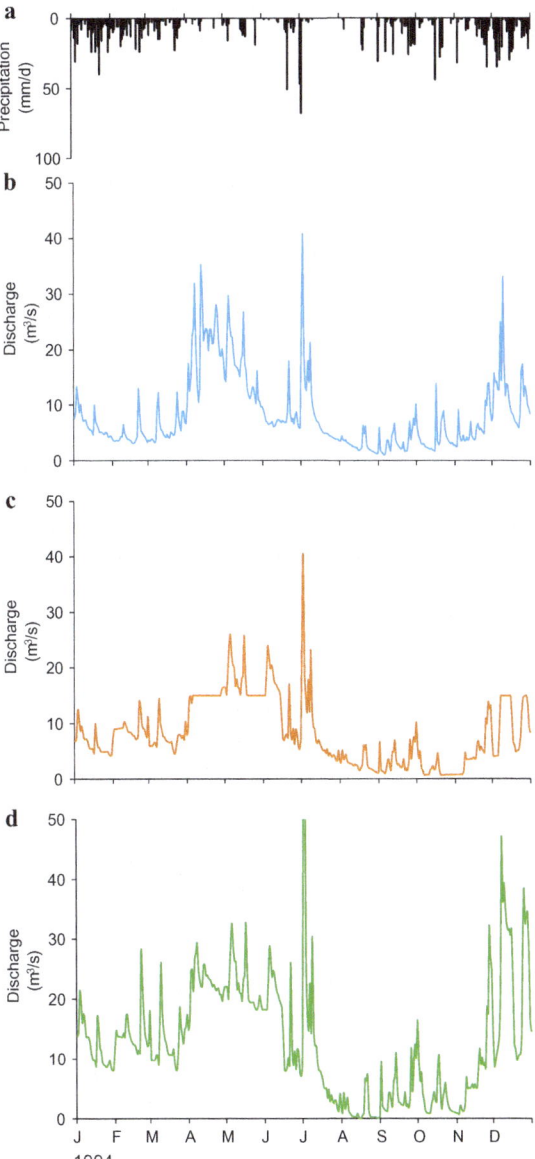

8.2.2 *Water Quality, Stone Surface Deposits and Macroinvertebrate Assemblage Survey*

The water quality, stone surface deposits and macroinvertebrate assemblage surveys
were conducted for approximately 20 days every month from April to December

1994. Furthermore, St. 1 could not be investigated in December 1994 because of snow accumulation. St. 1 to St. 4 were surveyed each time. The investigation start time at the St. 1 was at approximately 9:00, that at St. 2 was at approximately 11:00, that at St. 3 was at approximately 13:00 and that at St. 4 was at approximately 15:00.

At each survey site, water temperature, pH, DO, NO_3-N concentration, chlorophyll a concentration and sestonic organic matter concentration in the river water and chlorophyll a weight and organic matter weight on the stone surface deposits in the river were measured.

The water temperature was measured with a thermometer, avoiding direct sunlight. The pH was measured by using a colorimetric measuring instrument (manufactured by TOYO ROSHI KAISHA, LTD.), and the DO was measured by the Winkler method. In addition, water samples of 10 L were brought back to the laboratory in plastic bottles for the analysis of NO_3-N concentration, chlorophyll a concentration and sestonic organic matter concentration in the river water. These water samples were also used to analyze the weights of chlorophyll a weight and organic matter weight on the stone surface deposits in the river. Three fist-sized stones approximately 15 cm in diameter were collected at the rapids of the study sites, they were washed, and a square frame within 5 cm × 5 cm was scraped off of each stone with distilled water and a toothbrush.

The water was suctioned and filtered by using GF/C filter paper to measure the chlorophyll a concentration in the river water and the chlorophyll a weight on the stone surface deposits in the river. After extracting with acetone, the absorbance was measured by using a spectrophotometer (manufactured by Hitachi, Ltd., Model 200–10) and calculated by using the three-point method (UNESCO method). Furthermore, to measure the sestonic organic matter concentration in the river water and the organic matter weight on the stone surface deposits in the river. The water was suctioned and filtered by using previously weighed GF/C filter paper, and the filter paper was dried at 110 °C for 2 h and heated at 450 °C for 2 h. After releasing this cold filter, I weighed this filter and calculated the ignition loss. The chlorophyll a weight and organic matter weight on the stone surface deposits in the river were determined based on the average of the values for three samples at each survey site. The filtrate was frozen, and the NO_3-N concentration was measured by the hydrazine method.

In the rapids, 50 cm × 50 cm quadrats were randomly established to investigate the water quality and stone surface deposits. The quadrats were established at two places at each survey site. A dustpan-type wire mesh (opening 30 cm, depth 60 cm and height 30 cm) that was opened upstream was used to collect as much gravel and sand as possible, without collecting stones that would sink it. Macroinvertebrate assemblages in the gravel and sand were gently washed off in a plastic bucket with a saturated salt solution. The saturated salt solution was filtered by using gauze (mesh strong net, MS70). In addition, the macroinvertebrate assemblages were collected from the stone surface with tweezers. Then, the macroinvertebrate assemblages were fixed in 10% formalin solution.

The samples were brought back to the laboratory, sorted and identified by using a stereomicroscope according to Kawai (1985). First, the samples were classified into Trichoptera and other macroinvertebrate assemblages. In addition, 4-, 6-, 8-, 10-

and 12-month samples were classified as Ephemeroptera, Plecoptera, Diptera, Megaloptera and others. The Trichoptera were divided them into net-spinning caddisflies, Stenopsychidae, Hydropsychidae and others. The biomass of each order, Stenopsychidae and Hydropsychidae, was obtained by wet weight (g/m^2). Each individual macroinvertebrate assemblage was lightly touched with filter paper to remove surplus water. Each assemblage was weighed to 0.001 g per taxonomic group by using an electronic balance (manufactured by Mettler-Toledo International Inc., PM400). Individuals of Stenopsychidae were classified into *Stenopsyche marmorata* and *Stenopsyche sauteri* according to Aoya and Yokoyama (1987). However, it was not possible to classify the first instars of the two species, so they were grouped together as *Stenopsyche* sp. The number of individuals was calculated, and the population densities (No./m^2) of *Stenopsyche marmorata*, *Stenopsyche sauteri*, first instars of *Stenopsyche* sp. and Hydropsychidae were determined. The wet weight and population density of the macroinvertebrate assemblages at each survey site were estimated based on the mean value of the two samples.

8.3 Results

8.3.1 Water Quality and Stone Surface Deposits in the Ooishi River

The seasonal changes in water temperature at the survey sites are shown in Fig. 8.7. Water temperature indicated a tendency of being low to high from upstream to downstream each month. However, the water temperature difference among sites was large in summer and small in winter.

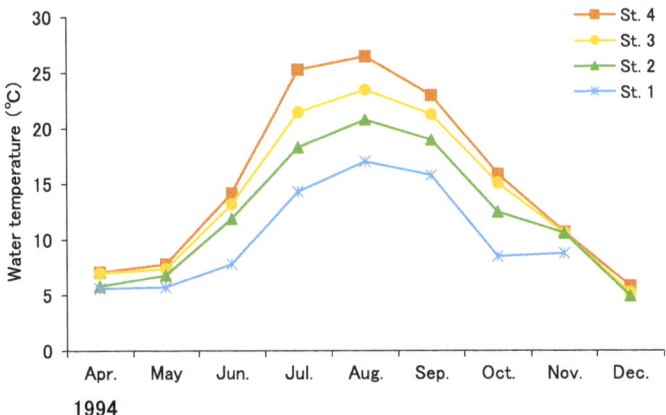

Fig. 8.7 Seasonal changes in water temperature at the survey sites. *Source* Reprinted from Matsui (2008). Copyright 2008 Ecology and Civil Engineering Society

Table 8.3 Water characteristics of the survey sites

Survey item	Upstream	Downstream		
	St. 1	St. 2	St. 3	St. 4
Water temperature (°C)	10.4 (4.6)	12.3 (6.0)	13.9 (6.9)	15.1 (8.1)
pH	7.0 (0.2)	6.9 (0.2)	7.0 (0.2)	7.1 (0.2)
DO (mg/L)	7.9 (0.9)	7.8 (1.1)	7.4 (1.1)	7.4 (1.3)
NO_3-N (μg/L)	219 (45)	207 (61)	211 (61)	240 (59)

Note Annual mean and SD (in parenthesis) are shown ($N = 9$)
Source Reprinted from Matsui (2008). Copyright 2008 Ecology and Civil Engineering Society

The annual mean and standard deviation of water temperature, pH, DO and NO_3-N at the survey sites are indicated in Table 8.3. Great differences in water quality were not observed among the survey sites.

Seasonal changes in chlorophyll a concentration and sestonic organic matter concentration in the river water at the survey sites are shown in Fig. 8.8. The chlorophyll a concentrations at downstream St. 2–4 were always greater than that at upstream St. 1 (1.4–10 times). Of the downstream sites, St. 2 had the greatest seasonal changes in chlorophyll a concentration and sestonic organic matter concentration in the river water. These values at St. 3 and St. 4 exceeded those at St. 2 from April to June. On the other hand, these values at St. 2 exceeded those at St. 3 and St. 4 from July to September. The sestonic organic matter concentrations at downstream St. 2–4 tended to be higher than that at upstream St. 1 except from April to May. Of the downstream sites, St. 2 experienced the greatest seasonal change in sestonic organic matter concentration, similar to the that in its chlorophyll a concentration. The sestonic organic matter concentration at St. 2 was high from June to September, and that at St. 4 was high in November.

Seasonal changes in chlorophyll a weight and organic matter weight on the stone surface deposits at survey sites are shown in Fig. 8.9. Chlorophyll a weights at downstream St. 2–4 showed a value greater than that at upstream St. 1 (2.3–12 times) from April to May, although the value downstream did not show a large difference from that upstream from summer to winter. The chlorophyll a weights at the downstream sites indicated a tendency to decrease in summer, and the change at St. 2 was generally substantial. The organic matter weights at downstream St. 2–4 were only 36–72% of that at upstream St. 1 from April to May, but the organic matter weights downstream were greater than that upstream from summer to winter. Among the downstream sites, the organic matter weight at St. 4 was lower than those at St. 2 and St. 3 from April to May. However, that at St. 4 rapidly increased in June and was significantly greater than those at St. 2 and St. 3 from summer to winter.

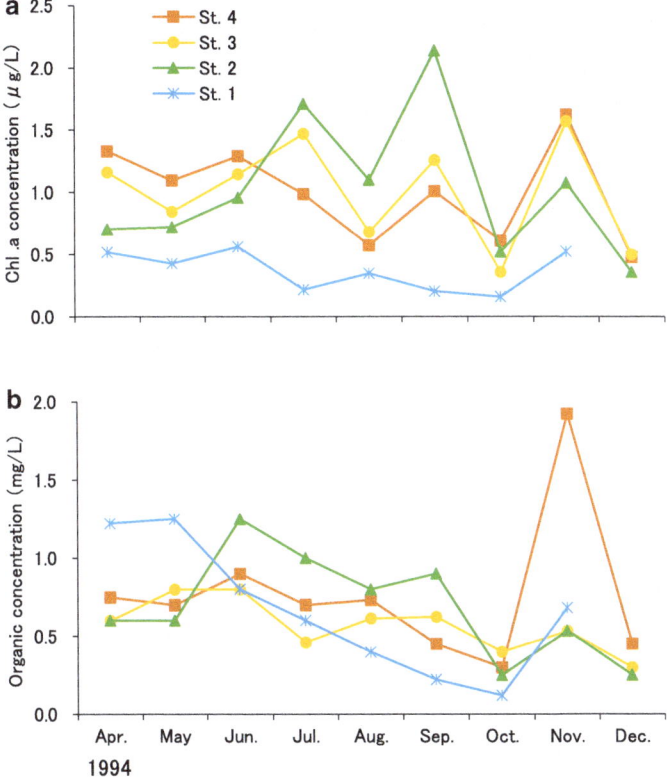

Fig. 8.8 Seasonal changes in concentrations of **a** chlorophyll a and **b** sestonic organic matter in the river water at the survey sites. *Source* Reprinted from Matsui (2008). Copyright 2008 Ecology and Civil Engineering Society

8.3.2 Macroinvertebrate Assemblages in the Ooishi River

The annual mean and standard deviation of macroinvertebrate assemblage biomass (wet weight) at each survey site are indicated in Table 8.4. Annual means of total biomass at downstream St. 2–4 were approximately 5–13 times that at upstream St. 1. However, the biomass of Trichoptera at St. 2–4 accounted for more than 74% of the total biomass, and that of Ephemeroptera and Plecoptera at St. 1 accounted for approximately 54% and 28% of the total biomass, respectively.

In terms of caddisflies, net-spinning caddisflies of Stenopsychidae at downstream St. 2–4 accounted for more than 98% of the total biomass, while those of Stenopsychidae and Hydropsychidae at upstream St. 1 accounted for only 47%.

Box plots and the statistical results for macroinvertebrate biomass at the survey sites are provided in Fig. 8.10 and Table 8.5. As a result of the Kruskal–Wallis test, total and Trichoptera biomass showed a significant difference. As a result of the

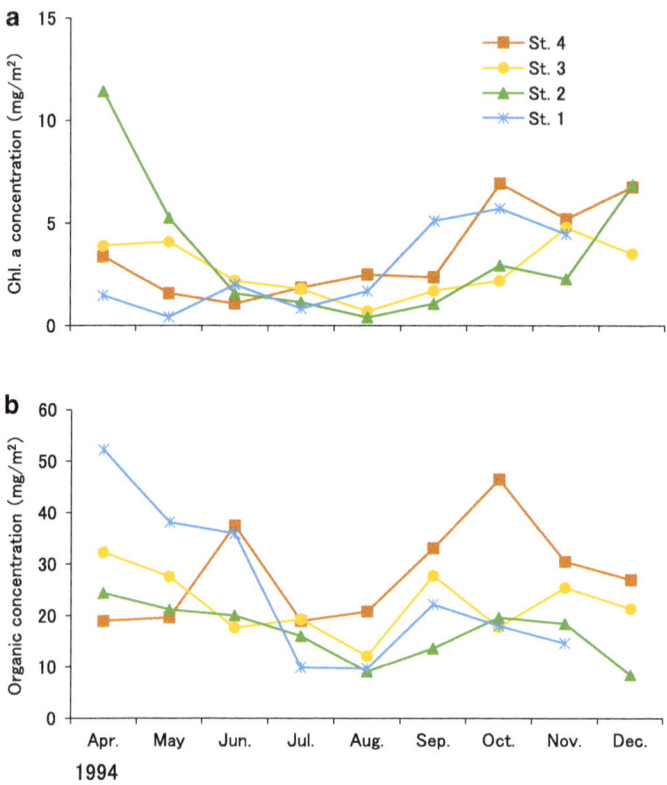

Fig. 8.9 Seasonal changes in stone surface deposits of **a** chlorophyll a and **b** organic matter at the survey sites. *Source* Reprinted from Matsui (2008). Copyright 2008 Ecology and Civil Engineering Society

Table 8.4 Macroinvertebrate biomass at the survey sites

Order	Upstream	Downstream		
	St. 1	St. 2	St. 3	St. 4
Ephemeroptera	0.56 (0.34)	0.21 (0.10)	0.57 (0.40)	0.47 (0.37)
Plecoptera	0.29 (0.53)	0.02 (0.05)	0.02 (0.05)	0.10 (0.08)
Trichoptera	0.07 (0.08)	4.07 (0.67)	10.16 (10.76)	11.65 (7.64)
Diptera	0.10 (0.13)	0.24 (0.26)	0.32 (0.23)	0.26 (0.16)
Megaloptera	–	0.89 (1.22)	0.08 (0.13)	0.56 (1.25)
Others	0.00 (0.01)	0.08 (0.17)	2.56 (4.62)	0.22 (0.30)
Total	1.03 (0.76)	5.51 (1.14)	13.71 (14.36)	13.26 (7.63)

Notes Wet weight (g/m^2). Annual mean and SD (in parenthesis) are shown ($N = 5$)
Source Reprinted from Matsui (2008). Copyright 2008 Ecology and Civil Engineering Society

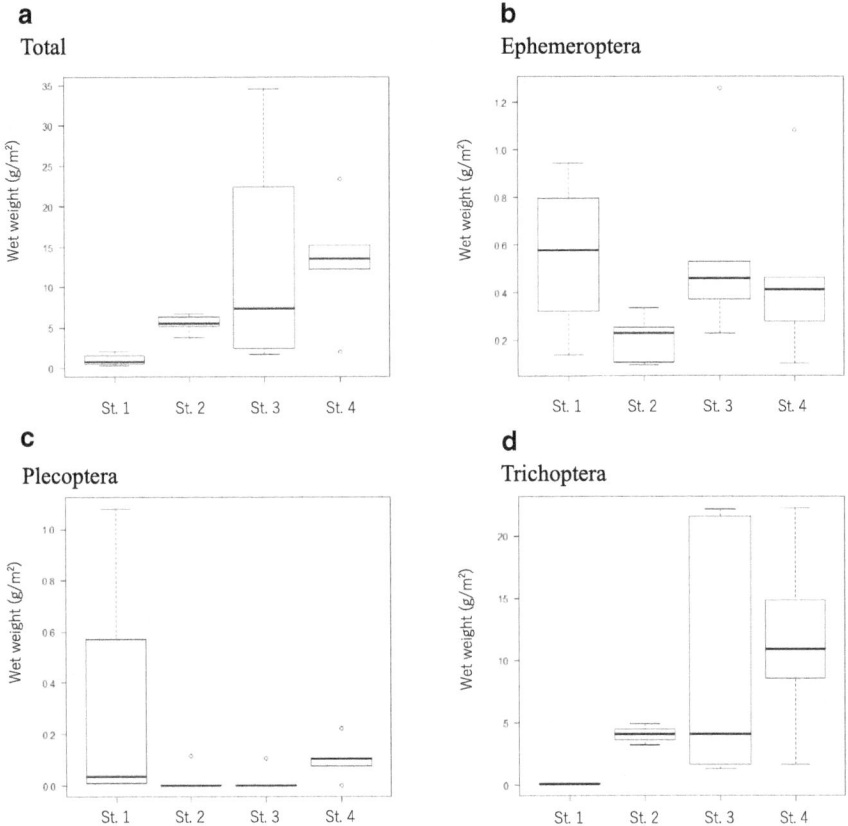

Fig. 8.10 Box plot of macroinvertebrate biomass at the survey sites. *Source* Illustration based on Matsui (2008)

Steel–Dwass test for total and Trichoptera biomass, no significant difference was confirmed.

8.3.3 Site Comparison of Taxon Density of Net-spinning Caddisflies

Illustrations of net-spinning caddisflies, Stenopsychidae and Hydropsychidae, are presented in Fig. 8.11. The seasonal changes in the densities of net-spinning caddisflies at the survey sites are shown in Fig. 8.12. The annual mean densities of *Stenopsyche marmorata, Stenopsyche sauteri,* first instar of *Stenopsyche* sp. and Hydropsychidae at downstream St. 2–4 were 7–32 times, 14–42 times, 1.9–3.5 times and 1.7–3.1 times greater than those at upstream St. 1, respectively. The annual mean

Table 8.5 Statistical results for macroinvertebrate biomass at the survey sites

a
Total
Kruskal–Wallis test P value = 0.031

Steel–Dwass test	St. 1	St. 2	St. 3	St. 4
St. 1		0.068	0.121	0.121
St. 2			0.953	0.397
St. 3				0.989
St. 4				

b
Ephemeroptera
Kruskal–Wallis test P value = 0.120

Steel–Dwass test	St. 1	St. 2	St. 3	St. 4
St. 1		0.315	0.994	0.883
St. 2			0.125	0.397
St. 3				0.953
St. 4				

c
Plecoptera
Kruskal–Wallis test P value = 0.165

Steel–Dwass test	St. 1	St. 2	St. 3	St. 4
St. 1		0.308	0.308	0.960
St. 2			0.999	0.455
St. 3				0.523
St. 4				

d
Trichoptera
Kruskal–Wallis test P value = 0.017

Steel–Dwass test	St. 1	St. 2	St. 3	St. 4
St. 1		0.068	0.068	0.068
St. 2			0.999	0.397
St. 3				0.884
St. 4				

Source Illustration based on Matsui (2008)

densities of *Stenopsyche marmorata* at St. 2 and St. 3 were 4–5 times that at St. 4. On the other hand, the annual mean density of *Stenopsyche sauteri* at St. 4 was 1.6–3 times those at St. 2 and St. 3. The annual mean density of the first instar of *Stenopsyche* sp. was very high at St. 2 from August to September. The annual mean density of Hydropsychidae did not show a noticeable difference among the downstream sites.

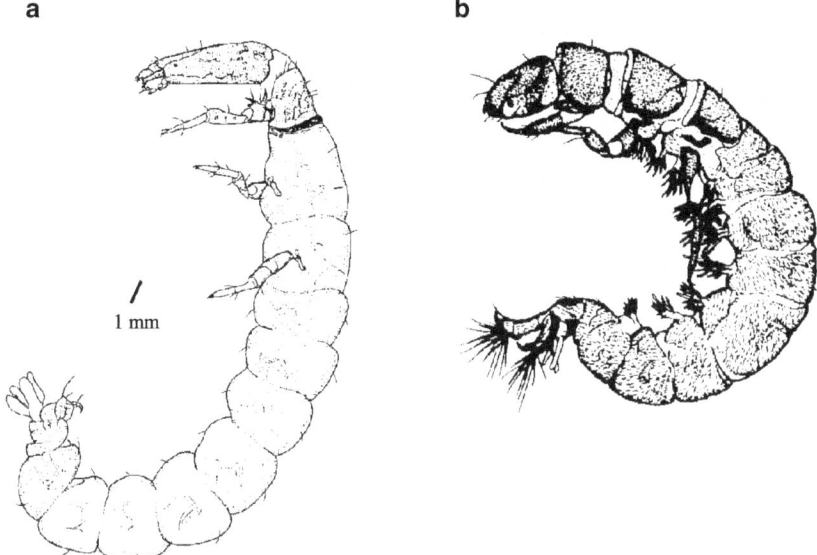

Fig. 8.11 Illustrations of net-spinning caddisflies: **a** Stenopsychidae and **b** Hydropsychidae. *Source* Reprinted from Kawai and Tanida (2005). Copyright 2005 Tokai University Publishing Division

8.4 Discussion

It was expected that the water temperature would increase at the sites downstream of the Ooishi Dam in summer because surface water is always being discharged. In this study, the temperature difference between the survey sites showed a tendency to be large in summer, although the temperature at St. 2 did not indicate a significantly large value. It was likely that the water temperature was not as high because the water at St. 2 consisted of inflow discharge from a tributary as well as outflow discharge from a dam.

The chlorophyll a and sestonic organic matter concentrations in the river water at the downstream sites were higher than those at the upstream site, with the exception of in spring. In addition, the chlorophyll a and sestonic organic matter concentrations in the river water at St. 2 increased in summer, and a high value was observed at St. 2 compared with those at the other sites. On the other hand, the chlorophyll a and organic matter weights on the stone surface deposits at the downstream sites were greater than those at the upstream site from summer to winter. However, the chlorophyll a and organic matter weights on the stone surface deposits at St. 2 did not increase in summer.

Based on the above information, the sestonic organic matter concentrations in the river water at the downstream sites were higher than that at the upstream site because the stone surface deposits peeled off and flowed to these sites. Thus, the increase in

Fig. 8.12 Seasonal changes in the densities of net-spinning caddisflies at the survey sites: **a** *Stenopsyche marmorata*, **b** *Stenopsyche sauteri*, **c** first instar of *Stenopsyche* sp. and **d** Hydropsychidae. *Source* Reprinted from Matsui (2008). Copyright 2008 Ecology and Civil Engineering Society

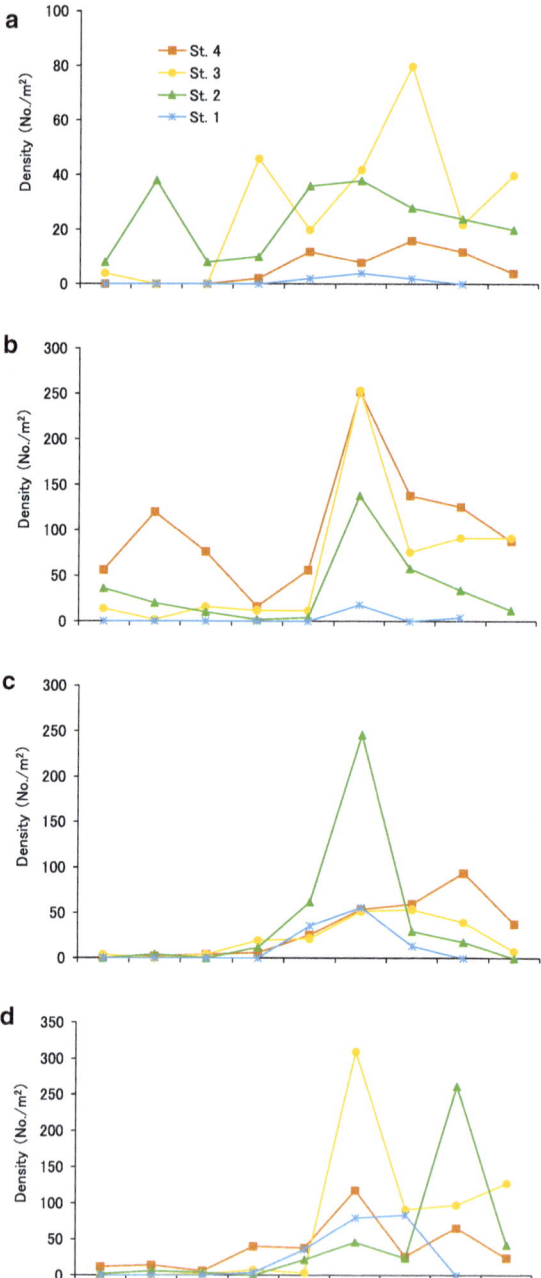

sestonic organic matter concentration in the river water at St. 2 in summer did not influence the peeling and flow of the stone surface deposits but was presumed to be due to the flow down of the phytoplankton grown in the Ooishi Dam reservoir.

The densities of the net-spinning caddisflies *Stenopsyche marmorata*, *Stenopsyche sauteri* and Hydropsychidae at the downstream sites were higher than those at the upstream site. The density of the first instar of *Stenopsyche* sp. was very high at St. 2 in summer. It was considered that the phytoplankton supplied from the Ooishi Dam reservoir became important food resources for juvenile larvae of *Stenopsyche* sp. Thus, the density of net-spinning caddisflies at the downstream sites was greater than that at the upstream site due to various factors related to the dam. In addition, the positive effect of the phytoplankton from the reservoir on the juvenile larvae of *Stenopsyche* sp. likely resulted in the increase in the number of individuals of net-spinning caddisflies at the downstream sites.

Thus, phytoplankton were supplied from the dam reservoir to the downstream river in summer, and it is likely that a new ecosystem was formed. It is important to develop a future dam control plan considering the significance of this ecosystem from various points of view. This river ecosystem where the number of individuals of *Stenopsyche* sp. is remarkable is not normal. A true river ecosystem consists of various species that live equally. When a river ecosystem becomes stable, the dominant life form of macroinvertebrate assemblages changes from the creeping type or case-bearing type to the net-spinning type (Mizuno and Gose 1993; Tsuda 1962) (Fig. 3.3). However, a natural ecosystem is established by being disturbed. When a dam reservoir is constructed in the upstream area of a river, a dam control plan that preserves the natural ecosystem as much as possible must be applied.

Box 8.1 Flexible Management of a Dam

Ministry of Land, Infrastructure, Transport and Tourism of Japan (2003) defined flexible dam management. Flexible management is implemented to preserve in downstream river environment in Japan. Dams store and discharge flowing water through the use of flood storage capacity. Flexible dam management is generally divided into 'increase discharge of the maintenance flow' and 'flash discharge' (Appendix Fig. 8.13).

Increasing the discharge of the maintenance flow means that the flow rate is increased above the maintenance flow rate of the operational regulation. In comparison with that of flash discharge, the flow rate is small and continually discharged. This discharge pattern is adopted to improve the habitat and environment of aquatic communities, landscapes, etc. On the other hand, flash discharge is a discharge used to temporarily cause a small-scale artificial flood, normalizing the water quality and ensuring the continuity of sand.

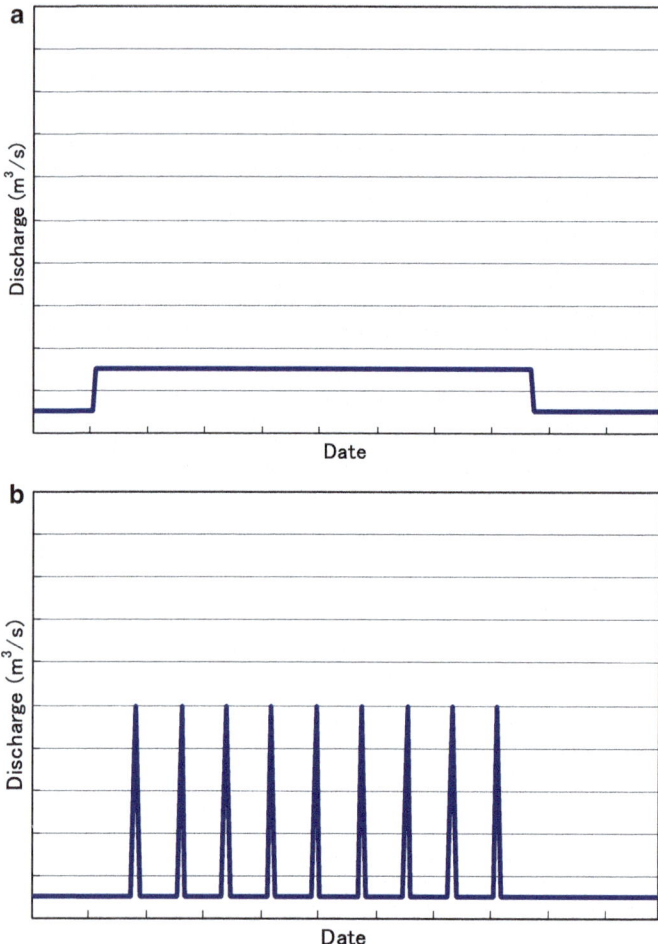

Fig. 8.13 Frame format of effluent pattern: **a** increase discharge of the maintenance flow and **b** flash discharge

Appendix

See Fig. 8.13.

References

Aoya K, Yokoyama N (1987) Life cycle of two species of Stenopsyche (Trichoptera: Stenopsy-chidae) in Tohoku District (in Japanese with English Abstract). Jpn J Limnol 48:41–53. https://doi.org/10.3739/rikusui.48.41

Furuya Y (1998) Downstream distribution and annual changes in densities of net-spinning Trichoptera (Hydropsychidae and Stenopsychidae) in the Yoshino River, Shikoku, Japan, with special reference to the colonization of *Macrostemum radiatum* Mclachlan (Trichoptera: Hydropsychidae) (in Japanese with English Abstract). Jpn J Limnol 59:429–441. https://doi.org/10.3739/rikusui.59.429

Kani T (1944) Ecology of mountain stream insects (in Japanese). In: Furukawa H (ed) Insects vol 1. Japanese Biological Magazine Volume 4. Kenkyusha Co., Lrd., Tokyo, pp 171–309

Kawai T, Tanida K (2005) Aquatic insects of Japan: manual with Keys and Illustrations (in Japanese). Tokai University Publishing Division, Kanagawa

Matsui A (2008) Effects of Oishi dam (Sekikawa Village, Niigata Prefecture, Japan) on downstream macroinvertebrate assemblages with particular focus on net-spinning caddisflies (in Japanese with English Abstract). Ecol Civ Eng 11:175–182. https://doi.org/10.3825/ece.11.175

Ministry of Land, Infrastructure, Transport and Tourism (2003) Guidance of flexible management testing of the dam (draft) (in Japanese). http://www.mlit.go.jp/river/shishin_guideline/dam5/pdf/danryokukanri_tebiki.pdf Accessed 25 June 2021

Mizuno N, Gose K (1993) Ecology of rivers (in Japanese). Tsukiji Shokan Publishing Co., Ltd., Tokyo

Morishita I (2001) Consideration for the conservation of ecosystem (in Japanese). Regional workshop transcription in fiscal 2001, The harmony with the environment in the agriculture and rural development project, Edited by The Japanese Institute of Irrigation and Drainage, 113–138, The Japanese Institute of Irrigation and Drainage, Tokyo

Okuma T (1999) Connection with the river and the people (in Japanese). A 10th anniversary symposium book of natural environment restoration association, Edited by Yokohama Environmental Science Research Institute, 92–95, Natural Environment Restoration Association, Shizuoka

Tanida K, Takemon Y (1999) Effects of dams on benthic animals in streams and rivers (in Japanese with English Abstract). Ecol Civ Eng 2:153–164. https://doi.org/10.3825/ece.2.153

Tsuda M (1962) Aquatic entomology (in Japanese). Hokuryukan and New Science Co., Ltd., Tokyo

Chapter 9
Lake Biwa Survey

Abstract The purpose of this study was to survey the distribution and habitat of submerged macrophytes in the Seta River outflow from Lake Biwa, Shiga Prefecture, Japan. To accomplish this, a field survey with sampling was conducted in January 2009. Nine submerged macrophyte species were collected during the diving operation. The sample collection showed that submerged macrophytes flourished on both sides of the river but did not flourish in the central portion of the river. Alien species (*Egeria densa* and *Elodea nuttallii*) were collected on all survey lines, while domestic species (*Ceratophyllum demersum*, *Myriophyllum spicatum*, *Hydrilla verticillata*, *Vallisneria biwaensis*, *Vallisneria denseserrulata*, *Potamogeton maackianus* and *Potamogeton malaianus*) were collected in large amounts on the survey lines near Lake Biwa. The dominant species in cohesive soil or fine soil was *Egeria densa*, while the dominant species in sand gravel was *Potamogeton maackianus*. Although *Egeria densa* was also collected in sand gravel, *Potamogeton maackianus* was not collected in cohesive soil or fine soil, which indicates that *Egeria densa* has high fertility and can survive in an oxygen-poor environment, while *Potamogeton maackianus* cannot survive in an oxygen-poor environment. A great amount of sunlight travels to the bottom of the lake because of artificial water-level manipulation in early summer. As a result, submerged macrophytes prosper in the summer. It is desirable to change the artificial water-level manipulation to control the extensive amount submerged macrophytes. For example, postponing the time for drawdown from 15 June to 15 July is preferable for not only controlling overgrowth of submerged macrophytes but also conserving fish spawning.

Keywords Alien species · Domestic species · *Egeria densa* · Lake Biwa · *Potamogeton maackianus* · Submerged macrophyte · Water-level manipulation

This chapter is a revised version of Matsui (2014). Copyright 2014 Springer: Springer. Landscape and Ecological Engineering. Relationship between distribution and bottom sediment of submerged macrophytes in the Seta River, Shiga Prefecture, Japan. Akira Matsui (2014), https://link.springer.com/article/10.1007%2Fs11355-012-0192-5, Accessed November 27, 2021.

© The Author(s), under exclusive license to Springer Nature Singapore Pte Ltd. 2022   115
A. Matsui, *Wetland Development in Paddy Fields and Disaster Management*,
https://doi.org/10.1007/978-981-19-3735-4_9

9.1   Introduction

The Seta River is the only river flowing from Lake Biwa. The Ministry of Construction (1987) reported that the Seta River was originally narrow and shallow. The water level of Lake Biwa used to rise and flood so frequently during times of high water because the discharge capacity of the Seta River was too low. As a result of the Seta River improvement project, the discharge capacity of the Seta River increased. Thus, the maximum water level of Lake Biwa has dropped dramatically.

The construction of a weir was completed in 1905. Since it was manually operated, it took a long time to open and close the weir. The present Arai Weir (see Fig. 9.1) was completed 120 m downstream from the old weir in 1961. The Lake Biwa Comprehensive Development Plan was conducted from 1972 to 1997, and after its implementation, the biomass of aquatic animals and plants in Lake Biwa changed.

For instance, Yamamoto et al. (2006) reported the probable effects of artificial water-level manipulations in Lake Biwa, initiated in 1992 to prevent flooding, on cyprinid fish larvae. The artificial reductions in the water level probably resulted in a significant decrease in the volume of shallow water in Lake Biwa and may be linked to the drastic decline in cyprinid fish.

Haga et al. (2006) noted that the amount of submerged macrophytes has increased in the southern basin of Lake Biwa since 1995. The distribution area of submerged macrophytes was approximately 27 km^2 from 1930 to 1940 and approximately 23 km^2 in 1953. However, a minimum of 0.6 km^2 was recorded in 1964, and between 1964 and 1994, the distribution area did not exceed 6 km^2. The area expanded to 9 km^2 in 1995, 16 km^2 in 1997, 29 km^2 in 2000 and 32 km^2 in 2001.

Overabundance of submerged macrophytes has also become a problem in the Seta River. Therefore, the distribution and habitat of submerged macrophytes were surveyed in the Seta River. On the basis of these results, the relationship between the distribution and bottom sediment of submerged macrophytes was determined. In addition, further reference to controlling the spread of submerged macrophytes is also included.

9.2   Methods

Sixteen survey lines were set in the Seta River, Shiga Prefecture, Japan, with six lines (St. R1-R6) along only the right bank, four lines (St. L1-L4) along only the left bank and six lines (St. 1-6) along both banks. St. R1-R6 and St. L1-L4 had one survey point in one survey line. St. 1-6 had three survey points in one survey line (Fig. 9.1). Submerged macrophytes and bottom sediment were surveyed at each survey point, and a field survey with sampling was conducted in January 2009.

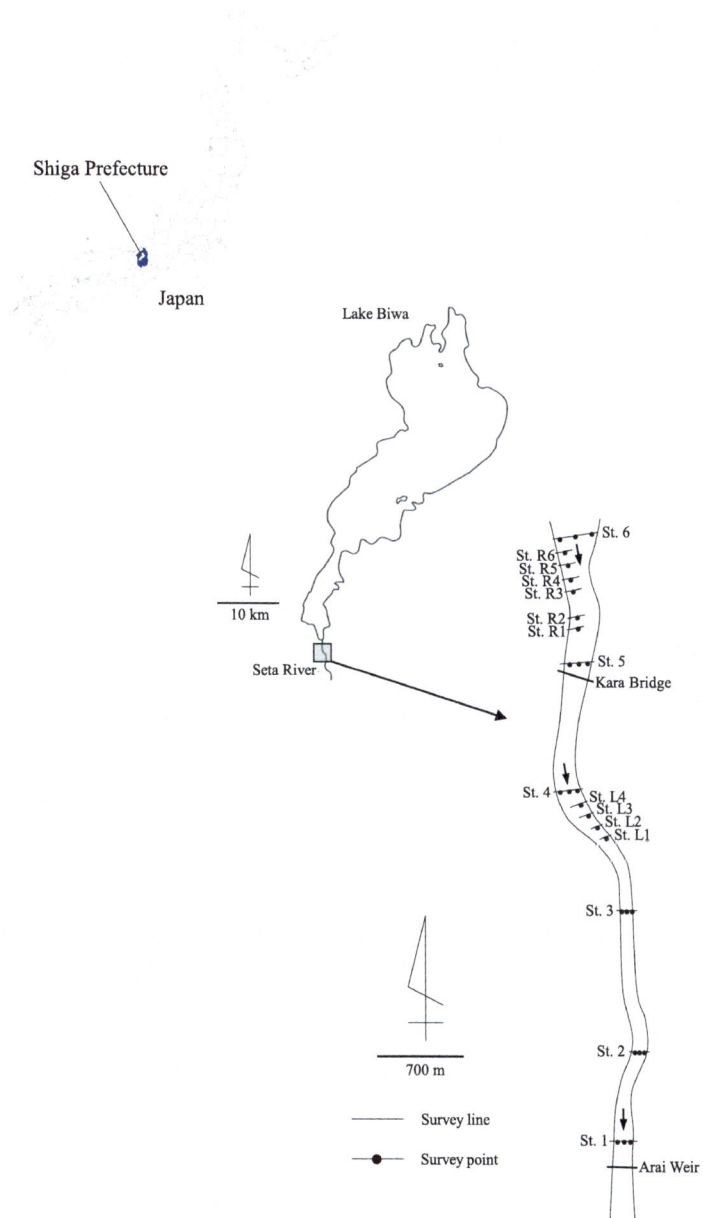

Fig. 9.1 Location of survey sites. *Source* Reprinted by permission from Springer: Springer. Landscape and Ecological Engineering. Relationship between distribution and bottom sediment of submerged macrophytes in the Seta River, Shiga Prefecture, Japan. Akira Matsui (2014)

For the submerged macrophyte survey, a visual observation during the diving operation was conducted using a belt transect method. The percentage of the river bottom covered by a plant community, the height of the plant community and the species name were determined.

The bottom sediment survey was carried out using a columnar bottom sampler. In the field, oxidation–reduction potential, soil color with revised standard soil color charts (Agriculture, Forestry and Fisheries Research Council 1986), soil smell and particle size by appearance were measured. In the laboratory, ignition loss, chemical oxygen demand, total nitrogen, total phosphorus, sulfide and total organic carbon of soils were measured. The analytical method for each item was based on the bottom sediment survey method (Environment Agency 1988).

9.3 Results

9.3.1 Submerged Macrophyte Survey

Nine submerged macrophyte species were collected during the diving operation (Table 9.1, Fig. 9.2). Submerged macrophytes were observed along the banks on both sides of the river, while there were few macrophytes along the center of the channel (Table 9.2). Two alien species, *Egeria densa* and *Elodea nuttallii*, were collected on all survey lines, while seven domestic species, *Ceratophyllum demersum, Myriophyllum spicatum, Hydrilla verticillata, Vallisneria biwaensis, Vallisneria denseserrulata, Potamogeton maackianus* and *Potamogeton malaianus*, were collected in large amounts on the survey lines near Lake Biwa (Table 9.2).

Table 9.1 Submerged macrophyte species collected in the survey area

Species	Domestic species or alien species?
Ceratophyllum demersum	Domestic species
Myriophyllum spicatum	Domestic species
Egeria densa	Alien species
Elodea nuttallii	Alien species
Hydrilla verticillata	Domestic species
Vallisneria biwaensis	Domestic species
Vallisneria denseserrulata	Domestic species
Potamogeton maackianus	Domestic species
Potamogeton malaianus	Domestic species

Source Reprinted by permission from Springer: Springer. Landscape and Ecological Engineering. Relationship between distribution and bottom sediment of submerged macrophytes in the Seta River, Shiga Prefecture, Japan. Akira Matsui (2014)

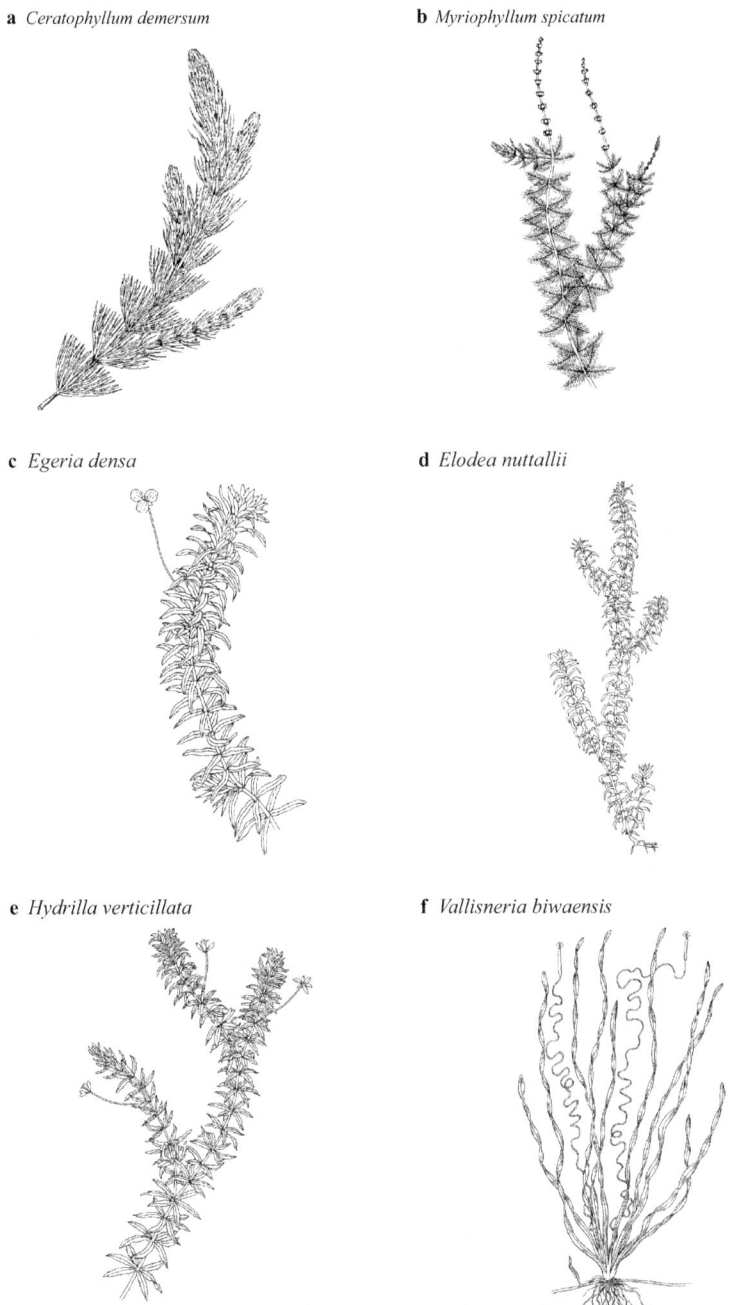

a *Ceratophyllum demersum*

b *Myriophyllum spicatum*

c *Egeria densa*

d *Elodea nuttallii*

e *Hydrilla verticillata*

f *Vallisneria biwaensis*

Fig. 9.2 Illustrations of submerged macrophyte species. *Source* Reprinted from Shiga Science Teaching Material Research Committee (1989). Copyright 1989

g *Vallisneria denseserrulata*

h *Potamogeton maackianus*

i *Potamogeton malaianus*

Fig. 9.2 (continued)

9.3.2   *Bottom Sediment Survey*

The soil colors of the bottom sediment samples collected at St. L1, St. L2, St. L3 and St. L4 were classified into blue or green yellow according to the revised standard soil color charts. The bottoms at these stations were covered with cohesive soil or fine soil (Table 9.2). Because of the fine sediments, oxygen was not supplied to the soil at these stations. As a result, the soil was in a reduced condition, and the soil had a metallic smell; thus, hydrated ferric oxide was likely dissolved in the soil. As the potential for oxidation reduction was negative and the sulfide content was high at these stations covered with silt and clay, it was clear that these stations were oxygen-poor environments.

The total nitrogen, total phosphorus, chemical oxygen demand, ignition loss, sulfide and total organic carbon concentrations of the soils in the samples at St.

Table 9.2 Water depth, bottom sediment and dominant species at each survey point

Survey line	Survey point	Water depth (m)	Bottom sediment	Dominant species
St. 1	Left	5.40	Sand gravel with cohesive soil	*Vallisneria denseserrulata*
	Center	5.50	Sand gravel	None
	Right	6.30	Sand gravel with cohesive soil	*Egeria densa*
St. 2	Left	4.20	Sand gravel	*Potamogeton maackianus*
	Center	7.30	Sand gravel with cohesive soil	None
	Right	4.60	Sand gravel	None
St. 3	Left	4.40	Sand gravel	*Potamogeton malaianus*
	Center	7.00	Sand gravel	None
	Right	6.30	Sand gravel	*Potamogeton maackianus*
St. 4	Left	1.55	Sand with silt	*Potamogeton maackianus*
	Center	6.40	Sand gravel	None
	Right	2.50	Sand gravel	*Potamogeton maackianus*
St. 5	Left	3.40	Sand gravel	*Egeria densa*
	Center	2.90	Sand gravel	*Hydrilla verticillata*
	Right	2.70	Sand gravel with cohesive soil	*Vallisneria denseserrulata*
St. 6	Left	2.50	Sand gravel	*Potamogeton malaianus*
	Center	6.50	Sand gravel	*Egeria densa*
	Right	1.10	Sand gravel	*Potamogeton maackianus*
St. L1	Left	2.80	Fine soil	*Egeria densa*
St. L2	Left	2.10	Silt with sand	*Egeria densa*
St. L3	Left	1.90	Sand with silt	*Egeria densa*
St. L4	Left	2.65	Sand	*Potamogeton maackianus*
St. R1	Right	2.70	Sand with silt	*Egeria densa*
St. R2	Right	2.85	Silt with sand	*Potamogeton maackianus*

(continued)

Table 9.2 (continued)

Survey line	Survey point	Water depth (m)	Bottom sediment	Dominant species
St. R3	Right	2.60	Fine soil	*Egeria densa*
St. R4	Right	3.35	Sand with silt	*Egeria densa*
St. R5	Right	2.25	Sand with gravel	*Egeria densa*
St. R6	Right	3.40	Fine soil	*Hydrilla verticillata*

Source Reprinted by permission from Springer: Springer. Landscape and Ecological Engineering. Relationship between distribution and bottom sediment of submerged macrophytes in the Seta River, Shiga Prefecture, Japan. Akira Matsui (2014)

L2, St. L3, St. R3 and St. 4 were relatively higher than those at the other stations (Table 9.3). The bottoms at these stations were covered with cohesive soil or fine soil, probably because of the low velocity of river flow in these areas, causing organic substances to be deposited there.

9.4 Discussion

9.4.1 Relationship Between Submerged Macrophytes and Bottom Sediment

The bottom sediment was cohesive soil or fine soil at 13 survey sites, while it was sand gravel at 15 survey sites (Table 9.4). Fifteen survey sites were on the right bank, center of flow and left bank, while 13 survey sites were generally not in the center of flow. As the flow velocity was large in the center of the river, cohesive soil or fine soil deposits were carried away by the flow.

The dominant species in the cohesive soil or fine soil was *Egeria densa*, and 64.2% of the area was covered by submerged macrophytes; the dominant species in the sand gravel was *Potamogeton maackianus*, and 39.2% of the area was covered by submerged macrophytes (Table 9.4).

Egeria densa was also collected in the sand gravel, but *Potamogeton maackianus* was not collected in the cohesive soil or fine soil. This result indicates that *Egeria densa* has high fertility and can survive in an oxygen-poor environment, while *Potamogeton maackianus* cannot survive in an oxygen-poor environment.

Table 9.3 Soil components at each survey point

Survey line	Survey point	Sulfide (mg/g)	Total nitrogen (mg/g)	Total phosphorus (mg/g)	Chemical oxygen demand (mg/g)	Ignition loss (%)	Total organic carbon (mg/g)
St. 1	Left	0.03	0.42	0.31	6.1	2.1	3.6
	Center	0.02	0.10	0.09	0.8	0.5	2.1
	Right	0.02	0.09	0.11	0.7	0.7	2.0
St. 2	Left	0.02	0.16	0.07	1.4	1.2	2.0
	Center	0.02	0.36	0.13	2.6	1.4	3.6
	Right	0.02	0.20	0.11	1.2	1.1	2.2
St. 3	Left	0.02	0.11	0.04	0.7	0.7	1.8
	Center	0.02	0.09	0.03	0.5	0.8	1.7
	Right	0.02	0.12	0.08	0.9	0.8	1.9
St. 4	Left	0.06	1.10	0.91	12.0	6.0	18.0
	Center	0.02	0.10	0.16	1.0	0.7	2.4
	Right	0.02	0.12	0.24	2.8	1.2	3.2
St. 5	Left	0.02	0.15	0.63	1.6	1.3	2.8
	Center	0.02	0.24	0.16	2.5	1.2	4.0
	Right	0.02	0.77	0.21	8.1	2.4	7.9
St. 6	Left	0.02	0.14	0.38	2.8	1.0	4.6
	Center	0.02	0.08	0.44	1.1	1.1	2.8
	Right	0.02	0.85	0.52	10.0	3.5	15.0
St. L1	Left	0.02	0.92	1.00	8.0	3.1	10.0
St. L2	Left	0.17	3.90	1.70	15.0	11.2	36.0
St. L3	Left	0.39	1.60	1.00	15.0	8.5	29.0
St. L4	Lett	0.04	0.86	0.74	8.8	3.9	12.0
St. R1	Right	0.02	0.09	0.12	3.7	1.3	3.9
St. R2	Right	0.06	0.94	0.23	10.0	4.3	16.0
St. R3	Right	0.11	1.30	1.00	17.0	8.4	37.0
St. R4	Right	0.12	0.47	0.45	7.3	2.4	7.7
St. R5	Right	0.02	0.28	0.70	8.3	3.3	9.9
St. R6	Right	0.02	0.59	0.51	5.4	2.2	13.0

Source Reprinted by permission from Springer: Springer. Landscape and Ecological Engineering. Relationship between distribution and bottom sediment of submerged macrophytes in the Seta River, Shiga Prefecture, Japan. Akira Matsui (2014)

Table 9.4 Relationship between bottom sediment and submerged macrophytes in the Seta River

Bottom sediment	Collection point		Percentage covered by plant community	Dominant species
Cohesive soil or fine soil	13	Left bank 5	64.2	*Egeria densa*
		Center of flow 1		
		Right bank 7		
Sand gravel	15	Left bank 5	39.2	*Potamogeton maackianus*
		Center of flow 5		
		Right bank 5		

Source Reprinted by permission from Springer: Springer. Landscape and Ecological Engineering. Relationship between distribution and bottom sediment of submerged macrophytes in the Seta River, Shiga Prefecture, Japan. Akira Matsui (2014)

9.4.2   Distribution Factor of Submerged Macrophytes

Haga et al. (2006) reported that *Potamogeton maackianus* was dominant, and *Hydrilla verticillata*, *Ceratophyllum demersum*, *Egeria densa* and *Myriophyllum spicatum* were also abundant in the southern basin of Lake Biwa in the summer of 2002. Therefore, it can be assumed that submerged macrophytes in the southern basin of Lake Biwa are also flourishing in the Seta River.

Based on the fact that the biomass of *Egeria densa* was negatively correlated with sediment diameter (Haga et al. 2006), *Egeria densa* utilizes cohesive soil or fine soil as a habitat. As *Egeria densa* is a highly competitive invader (Yarrow et al. 2009), *Egeria densa* will be able to inhabit environments under adverse conditions where native species cannot survive.

On the other hand, the biomass of *Potamogeton maackianus* was positively correlated with the average transparency/water depth ratio, suggesting that *Potamogeton maackianus* tends to grow in bottom areas under strong light conditions (Haga et al. 2006). As *Potamogeton maackianus* was the dominant species in sand gravel in this survey, the distributional factor of *Potamogeton maackianus* may involve both the light conditions and bottom sediment conditions.

9.4.3 Influence of Artificial Reductions in Water Level on Submerged Macrophytes

Water levels in Lake Biwa are regulated by the Arai Weir, which was constructed 5 km downstream in the Seta River to control the water supply and prevent flooding (Yamamoto et al. 2006). This artificial water-level manipulation is used to create a drawdown of 20–30 cm from 15 June to 15 October (Fig. 9.3). A great amount of sunlight travels to the bottom of the lake because of this manipulation. As a result, submerged macrophytes prosper in the summer. This water-level manipulation began in 1992, and at the same time, the expansion of submerged macrophytes began.

Therefore, it is necessary to change this artificial water-level manipulation to control the overgrowth of submerged macrophytes. For example, postponing the drawdown time from 15 June to 15 July is preferable for not only controlling the overgrowth of submerged macrophytes in early summer but also conserving fish spawning. When submerged macrophytes are present in adequate amounts, the flow velocity in the Seta River is high. The bottom sediment under submerged macrophytes is refreshed by the flow, and consequently, the biomass of submerged macrophytes will be properly maintained.

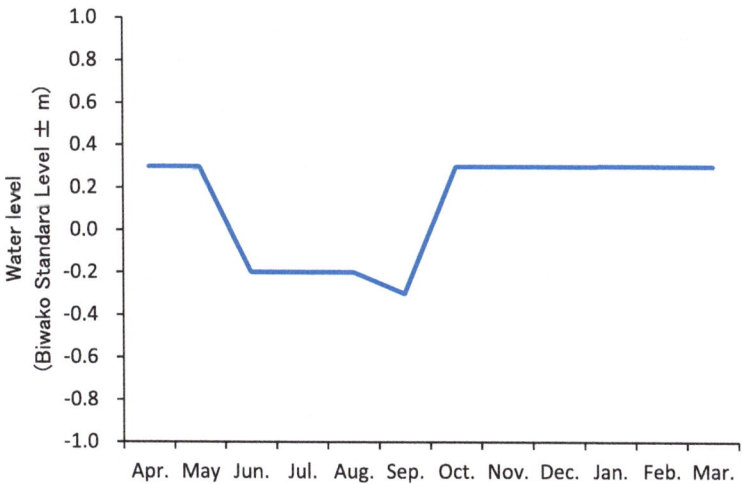

Fig. 9.3 Water-level operations in the Lake Biwa. *Source* Modified from the Incorporated Administrative Agency Japan Water Agency Lake Biwa Development Integrated Operation & Maintenance Office, https://www.water.go.jp/kansai/biwako/, Accessed November 1, 2021 (in Japanese)

Fig. 9.4 Uprooted removal of submerged macrophytes

Fig. 9.5 Surface mowing of submerged macrophytes

Box 9.1 Overgrowth and Removal Measures of Submerged Macrophytes in the Southern Basin of Lake Biwa

Shiga Prefecture (2013) reported submerged macrophyte overgrowth in the southern basin of Lake Biwa and provided removal measures. Submerged macrophytes in the southern basin of the lake suddenly increased during a drought in 1994, and since that time, the lake has been under the abnormal

Fig. 9.6   Manual mowing of submerged macrophytes

Fig. 9.7   Young shells in Lake Biwa

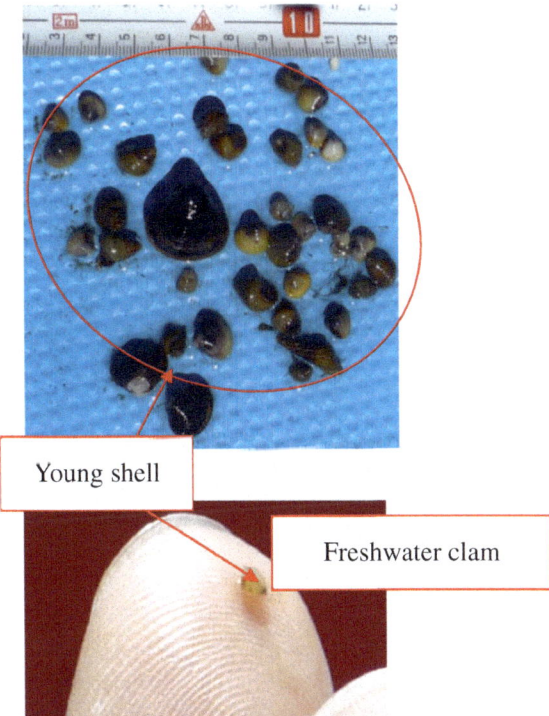

condition of submerged macrophytes covering approximately 90% of the bottom of the lake in summer. Overgrowth of submerged macrophytes causes problems in the natural environment and ecological systems by slowing lake currents, and these problems can include the deterioration of water quality, low-oxygen water masses in the bottom layers and sludge at the bottoms of lakes. In addition, the overgrowth of submerged macrophytes has led to various issues for humans as they can become obstacles to fishing and navigating ships and have putrid odors.

Submerged macrophyte removal measures have involved two methods: uprooted removal and surface mowing. Uprooted removal of submerged macrophytes has been implemented over hundreds of meters in width in the north–south direction of the basin with fishing and shellfish boats dragging fishing gear to remove the macrophytes and restore lake flow. Buoys have been set up in the range of 400 × 500 m, and 40 fishing boats have sailed from upstream to downstream (Appendix Fig. 9.4). In addition, submerged macrophyte habitat on the surface layer (1.5 m water depth) has been mowed with a special ship that has excellent maneuverability (Appendix Fig. 9.5). As emergency job creation promotional business, surface mowing has also been performed manually at shallow water sites where a ship cannot go (Appendix Fig. 9.6).

As a result, overgrowth of new submerged macrophytes has been suppressed in the watershed where uprooted removal of submerged macrophytes has been implemented over multiple years. In addition, the lake bottom environment has been improving as the poor oxygen zone has decreased in size and young shells have been restored (Appendix Fig. 9.7).

Appendix

See Figs. 9.4, 9.5, 9.6 and 9.7.

References

Agriculture, Forestry and Fisheries Research Council (1986) Revised standard soil color charts (in Japanese). Japan Color Enterprise Co.Ltd., Tokyo

Environment Agency (1988) Bottom sediment survey method (in Japanese). Environment Agency, Tokyo

Haga H, Ohtsuka T, Matsuda M, Ashiya M (2006) Spatial distributions of biomass and species composition in submerged macrophytes in the southern basin of Lake Biwa in summer of 2002 (in Japanese with English Abstract). Jpn J Limnol 67:69–79. https://doi.org/10.3739/rikusui. 67.69

Matsui A (2014) Relationship between distribution and bottom sediment of submerged macrophytes in the Seta River, Shiga Prefecture, Japan. Landscape Ecol Eng 10:109–113. https://doi.org/10. 1007/s11355-012-0192-5

Ministry of Construction (1987) Seta River Weir. Ministry of Construction, Tokyo

Shiga Prefecture (2013) Waterweed measures business in Lake Biwa (in Japanese). https:// www.pref.shiga.lg.jp/kensei/gaiyou/soshiki/biwakokankyoubu/biwakohozensaiseika/. Accessed 1 Nov 2021

Shiga Science Teaching Material Research Committee (1989) Aquatic plants of shiga, illustrated handbook (in Japanese). Shingakusha Co.Ltd., Kyoto

Yamamoto T, Kohmatsu Y, Yuma M (2006) Effects of summer drawdown on cyprinid fish larvae in Lake Biwa. Jpn Soc Limnology 7:75–82. https://doi.org/10.1007/s10201-006-0172-2

Matthew Y, Victor MH, Max F, Antonio T, Luisa DE, Fernanda F (2009) The ecology of *Egeria densa* Planchon (Liliopsida: Alismatales): a wetland ecosystem engineer? Rev Chil Hist Nat 82:299–313. https://doi.org/10.4067/S0716-078X2009000200010

Part IV
Case Study of Paddy Fields

Chapter 10
Drainage Canal System Survey

Abstract The importance of paddy field ecosystems is being emphasized in consideration for biodiversity. However, we do not know what kind of paddy ecosystem is most effective for biodiversity. To research this problem, I investigated the distribution and life history of aquatic animals in a consolidated paddy field. I surveyed various aspects of the paddy field, such as canal levels with the main, lateral and farm drains, areas with the presence or absence of a year-round water flow, as well as sections with different canal bed materials. These findings provide knowledge for determining an agricultural and rural development project that can be harmonized with the environment. In this study, I selected six survey sites for drainage canals in Shimodate City (now Chikusei City), Ibaraki Prefecture, Japan. The sampling interval was one month from April 2001 to March 2002. A survey of fishes revealed that *Opsariichthys platypus* and *Misgurnus anguillicaudatus* were concentrated in the main drains and the lateral and farm drains, respectively. Among aquatic insects, damselfly *Calopteryx atlata* and dragonfly *Orthetrum albistyrum speciosum* were observed in the lateral drains and the farm drains, respectively. Thus, the drainage canal system was believed to function as a spawning ground and habitat for four species. It is especially important for four species to have a year-round water flow and natural materials for canal beds. To enrich biodiversity in consolidated paddy fields, it is effective to create a wetland as a wintering site for four species.

Keywords Agricultural and rural development project · Aquatic animal · Biodiversity · Distribution · Drainage canal · Farmland consolidation · Growth · Paddy field · Wetland · Wintering site · Year-round water flow

10.1 Introduction

Conventional paddy farmland consolidation in Japan, which aims to increase farming efficiency by improving the drainage conditions of paddy fields and independently creating irrigation and drainage canals, has negative impacts on biodiversity in rural

This chapter is a revised version of Matsui and Satoh (2004a). Copyright 2004a, https://doi.org/10.18960/hozen.9.2_153, Accessed November 16, 2021; and Matsui (2009). Copyright 2009, https://doi.org/10.18960/hozen.14.1_3, Accessed November 16, 2021.

© The Author(s), under exclusive license to Springer Nature Singapore Pte Ltd. 2022   133
A. Matsui, *Wetland Development in Paddy Fields and Disaster Management*,
https://doi.org/10.1007/978-981-19-3735-4_10

areas (Arai 2001; Food, Agriculture and Rural Policy Council 2002; Fujioka 1998; Hasegawa 1998; Hata 1987; Moriyama 1997; Nakagawa 2000; Ozawa 2000; Ueda 1998).

The Land Improvement Act of Japan was amended in June 2001 and requires agricultural and rural development projects be environmentally sustainable. It is widely recognized that transforming concrete irrigation and drainage canals into earthen canals and minimizing the differences in elevation between paddy plots and drainage canals aid in the preservation of aquatic animals. However, most paddy fields that have been consolidated using conventional standards will inevitably remain intact, thus continuing to have a substantial influence on the regional environment. Therefore, at least the minimum environmental measures should be effectively implemented in consolidated paddy fields. To this end, the distribution of the aquatic animals associated with these fields requires clarification. In addition, this knowledge will be useful in determining how to effectively distribute water during a nonirrigation season.

The agricultural waterways created by paddy farmland consolidation have been functionally divided into irrigation and drainage canals, and these have been further divided based on location into main, lateral and periphery. Aquatic animals appear to have selected their habitats in response to the environmental conditions in each of these types of waterways. To manage these agricultural waterways considering environmental aspects in the future, determining aquatic animal habitats in each type of canal is essential.

In general, irrigation canals are mainly concrete, three-sided, lined waterways used for efficient water supply. Given that there is no water in irrigation canals in the nonirrigation season, aquatic animal conservation must be incorporated into drainage canals. Hata (1987) described that the consistent development of fish habitat in both irrigation and drainage canal systems is ideal, but as a general theory, it is realistic that the main target of fish conservation is the drainage canal system.

The purpose of this study was to clarify the distribution and growth of aquatic animals in the canal systems of the main, lateral and farm drains in consolidated paddy fields with a special focus on canal structure and year-round water flow in the canals. A field survey at six sites, which were selected for their different canal levels, was carried out in Shimodate City (now Chikusei City), Ibaraki Prefecture, Japan (36°21'N, 139°59'E). The sampling interval was one month from April 2001 to March 2002.

Furthermore, based on the concept of biodiversity conservation in this area, it is important to focus on preserving the environment that representative aquatic animals of the regions inhabit rather than on ensuring the number of species is large. If agricultural and rural development projects are implemented within a range that can satisfy the above conditions, then compatibility between agricultural productivity improvement and regional ecosystem conservation will be possible.

10.2 Methods

10.2.1 Survey Areas

The study sites were located in the drainage canal under the jurisdiction of Kawama Land Improvement District, Shimodate City (now Chikusei City), Ibaraki Prefecture, Japan. The area of the Kawama Land Improvement District is located in the northeast in Shimodate City. The northern end of this area touches Tochigi Prefecture, and the east and west are located between the Kokai River and Gogyo River. This area is located in the paddy field zone and forms a slender area in which the distance to the east and west is approximately 2.5 km and that to the north and south is approximately 9 km. Villages are scattered within this area. The drainage canal receives drainage water from the irrigation area, and this is taken from 2 points: the Akaido diversion weir in the Gogyo River and the Ochiai diversion weir in the left tributary in the Gogyo River. Drainage water drains naturally into the Gogyo River via the farm drain, lateral drain and main drain.

The survey sites were located among the abovementioned drainage systems and considered the presence of running water for the nonirrigation season. Generally, six sites were set up; two sites in the main drain (MD1 and MD2), two sites in the lateral drain (LD1 and LD2) and two sites in the farm drain (FD1 and FD2) (Table 10.1; Figs. 10.1, 10.2, 10.3, 10.4, 10.5, 10.6 and 10.7).

The irrigation season in this area is from the middle of April to the beginning of September. It is difficult to clearly sort the forms because the main and lateral

Table 10.1 Basic features of the survey sites

Survey sites		Year-round water flow	Bottom material	Side material	Width (m)	Depth (m)	Length (m)	Consolidation (year)
MD1	Main drain 1	○	Concrete	Concrete	3.20	1.20	80	1967
MD2	Main drain 2	○	Gravel and sand	Concrete	2.71	1.20	100	1964
LD1	Lateral drain 1	○	Gravel and sand	Concrete	2.22	1.20	100	1964
LD2	Lateral drain 2	×	Gravel and sand	Concrete	1.42	1.20	100	1964
FD1	Farm drain 1	○	Sand and mud	Concrete	0.78	0.90	135	1964
FD2	Farm drain 2	×	Sand and mud	Concrete	0.78	0.90	135	1964

Note Width: canal width, depth: canal depth and length: surveyed length
Source Reprinted from Matsui and Satoh (2004a). Copyright 2004a

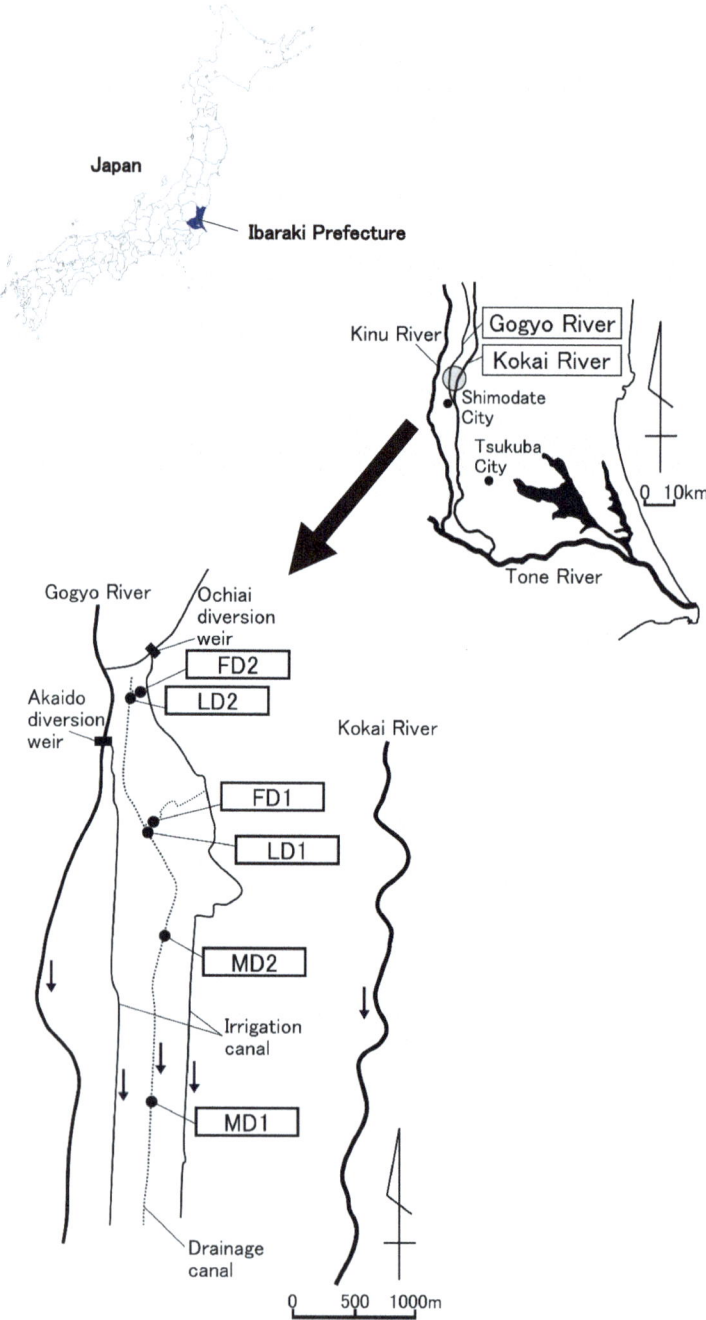

Fig. 10.1 Location of the survey sites. *Note* For information on the survey sites MD1, MD2, LD1, LD2, FD1 and FD2, see Table 10.1. *Source* Reprinted from Matsui and Satoh (2004a). Copyright 2004a

Fig. 10.2 Panoramic view at main drain 1 (*Photo* by Akira Matsui). *Notes* Photo date: July 2001. View from downstream to upstream

Fig. 10.3 Panoramic view at main drain 2 (*Photo* by Akira Matsui). *Notes* Photo date: July 2001. View from downstream to upstream

Fig. 10.4 Panoramic view at lateral drain 1 (*Photo* by Akira Matsui). *Notes* Photo date: July 2001. View from downstream to upstream

Fig. 10.5 Panoramic view at lateral drain 2 (*Photo* by Akira Matsui). *Notes* Photo date: July 2001. View from downstream to upstream

Fig. 10.6 Panoramic view at farm drain 1 (*Photo* by Akira Matsui). *Notes* Photo date: July 2001.
View from downstream to upstream

Fig. 10.7 Panoramic view at farm drain 2 (*Photo* by Akira Matsui). *Notes* Photo date: July 2001.
View from downstream to upstream

drains do not branch morphologically. Based on the differences in the physical environmental conditions of water depth and current velocity, two stations on the downstream side were considered the main drains, and two stations on the upstream side were considered the lateral drains. For the canal structure, the sidewall material is concrete at all survey sites, and the bed material is as follows: MD1 has concrete bed material; MD2, LD1 and LD2 have gravel and sand bed material; and FD1 and FD2 have sand and mud bed material. There is year-round water flow at MD1, MD2, LD1 and FD1, while there is no running water during the nonirrigation season at LD2 and FD2. A miscellaneous drainage from villages around FD1 maintains the running water. Additionally, there is an elevation of approximately 0.30 m at the intersection of LD1 and FD1 and at the intersection of LD2 and FD2.

The consolidation of paddy farmland in this area started as an agricultural structure improvement project in 1964 and was completed in 1975. Before maintenance, this area had an abundant quantity of water with ample water for irrigation. However, drainage was poor, and farm roads were not constructed. This area was an important paddy field area in Ibaraki Prefecture; however, it was in an unstable farming state. Five stations except for MD1 were established in 1964, and MD1 was established in 1967. The agricultural waterways created by paddy farmland consolidation were functionally divided into irrigation and drainage canals, but they were all open channels that were not yet used for the purpose of thoroughly drying paddy fields, which enabled crop rotations.

The water level is raised by the Akaido diversion weir and irrigation water obtained by operating the gate of the water intake gutter pipe on the left bank. As there is no large difference in elevation from the Akaido diversion weir to the paddy field, fish can move freely. On the other hand, as the drainage water flows from the paddy field to the Gogyo River, there is a difference in elevation of approximately 1 m between the paddy field and the farm drain and approximately 3 m between the main drain and Gogyo River; thus, fish cannot move between the paddy field and the drainage canal system and between the drainage canal system and the Gogyo River. An upstream gate was installed in the main drain just before the Gogyo River to ensure a constant water level. The gate is dropped to reuse the drainage water in this area as the irrigation water in the downstream area during the irrigation season.

10.2.2 Survey Methods

Physical environment and aquatic animal surveys in the drainage canal system were carried out once a month from April 2001 to March 2002 on the 20th day of each month. The surveys were conducted on days with good weather as much as possible. The surveys took two days; on the first day, surveys were performed at MD1, MD2 and LD1, and on the second day, surveys were carried out at FD1, LD2 and FD2. Surveys took place from early morning to sunset and from downstream to upstream. In addition, because there was no running water in the drainage canal system, LD2 and FD2 were not investigated from January to April and September to April, respectively.

10.2.3 Physical Environmental Survey

Water depth, flow velocity, water temperature, hydrogen ion concentration (hereafter called the pH), dissolved oxygen (DO) concentration and electrical conductivity (EC) were measured. Water depth was measured with a meter stick. Flow velocity was measured with a waterproof separation flow velocity meter (manufactured by Miura Rika Industries, Ltd., Model CR-7 WP). Water temperature, pH, DO and EC were determined by a water quality meter (manufactured by DKK-TOA Corporation, Model WQC-22A). The water depth and the flow velocity were determined based on the measure of a point at the center of FD1 and FD2. These values were a mean of three measures, which were divided by four into the crossing direction at MD1, MD2, LD1 and LD2. Incidentally, the flow velocity was measured at a 60% water depth point. A self-recording pressure-type water-level gauge (manufactured by TOKYO KEISOKU CO., LTD., data logger model HG-4101, pressure sensor model HGT-01) was installed at LD1 near the center of this study area, and the water level was recorded continuously every hour.

10.2.4 Aquatic Animal Survey

A fish survey was carried out by using a cast net (diameter 2 m, net height 3 m and opening size 12 mm) and a spoon net (bottom 0.35 m, net height 0.30 m and opening size 3 mm). The fish survey with the cast net was conducted five times at each site in the main and lateral drains. In addition, the fish harvested with the spoon net were swept by the net at an interval of two meters along both sidewall portions of the surveyed length at all sites.

The caught fish, except for large fish, were immediately fixed with a 10% formalin aqueous solution in the field. After returning to the laboratory, the species of fishes were identified according to Nakabo (2000), and the standard length and wet weight were measured. The wet weight was measured with an electronic balance (manufactured by Sartorius Japan K. K., Model BP1200). The large fish were released after the standard length and wet weight were measured in the field. It should be noted that the minimum units of the standard length and wet weight were 0.1 cm and 0.01 g, respectively.

In addition, an aquatic insect survey was carried out by using a spoon net (described above). Aquatic insects harvested with the spoon net were swept by the net at an interval of two meters along both sidewall portions of the surveyed length at all sites.

The caught aquatic insects were immediately fixed with a 10% formalin aqueous solution in the field. After returning to the laboratory, the species of aquatic insects were identified according to Kawai (1985) and Ishida et al. (1988) by using a stereomicroscope (manufactured by Carton Optical Industries, Ltd., Model NSZT-44E). The number of individuals and wet weight were measured. The wet weight was determined with an electronic balance (manufactured by Mettler-Toledo International Inc., PM400). The head widths of *Orthetrum albistylum speciosum* and *Calopteryx atrata*, which were caught many times, were determined. It should be noted that the minimum units of the head width and wet weight were 0.05 mm and 0.001 g, respectively.

10.3   Results

10.3.1   Physical Environment in the Drainage Canal System

As factors affecting the hydrological conditions of the survey area, seasonal changes in precipitation and water depth at LD1 are shown in Fig. 10.8. The precipitation was recorded by the Automated Meteorological Data Acquisition System (AMeDAS) in Kyowa Town near Shimodate City. The total precipitation from April 2001 to March 2002 was 1116 mm, which was close to the average value of 1167 mm over the past 20 years. The water depth was generally high during the irrigation season from May to August and fluctuated greatly according to rainfall. The water depth at LD1 increased to 1.00 m in August and October due to flooding.

Seasonal changes in water depth and flow velocity at the survey sites are presented in Fig. 10.9. In general, the water depth and flow velocity were greater in the irrigation season at all survey sites, and they showed a minimal trend in the nonirrigation season.

The seasonal changes in DO and EC at the survey sites are presented in Fig. 10.10. The DO at FD1 was remarkably low from September to April during the nonirrigation season. It tended to fall below 5 mg/l, which is a water quality standard of agricultural water (The Japanese Society of Irrigation 2000). The EC at FD1 was remarkably high from October to April during the nonirrigation season. It tended to exceed 30 mS/m, which is the water quality standard of agricultural water (The Japanese Society of Irrigation 2000).

The seasonal changes in water temperature and pH at the survey sites are shown in Fig. 10.11. The maximum water temperature at all survey sites occurred in July 2001. The minimum water temperature at MD1 and MD2 occurred in January 2002; that at LD1, LD2 and FD1 occurred in December 2001; and that at FD2 occurred in June 2001 because no water flows in winter. In general, the pH at all survey sites ranged from 6.0 to 7.5, which meets the water quality standard of agricultural water (The Japanese Society of Irrigation 2000).

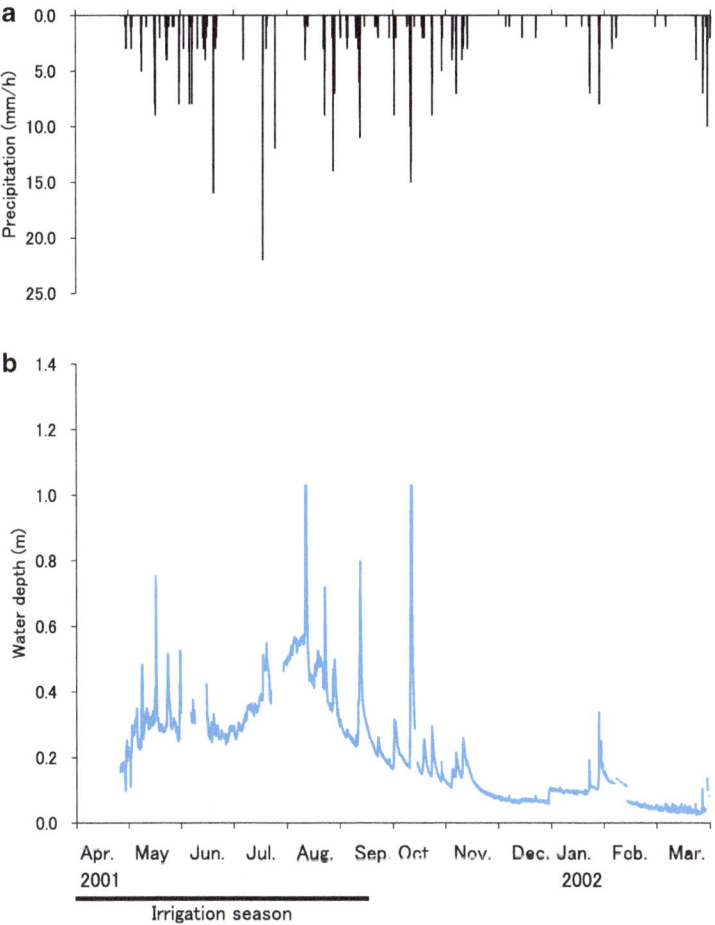

Fig. 10.8 Seasonal changes in **a** precipitation and **b** water depth at lateral drain 1. *Source* Reprinted from Matsui and Satoh (2004a). Copyright 2004a

10.3.2 Total Numbers of Fish Caught at the Survey Sites

In total, 694 individual fish were caught at all survey sites. The species were as follows: *Cyprinus carpio*, *Carassius* sp., *Opsariichthys platypus*, *Candidia temminckii*, *Tribolodon hakonensis*, *Pseudorasbora parva*, *Gnathopogon elongatus elongatus*, *Pseudogobio esocinus esocinus*, *Misgurnus anguillicaudatus*, *Silurus asotus*, *Oryzias latipes* and *Rhinogobius* sp. (Table 10.2). The values in the table show the actual number of fish caught at each of the survey sites. The order of listed fishes follows Nakabo (2000). The identification of *Candidia temminckii* was based on Hosoya et al. (2003). All *Misgurnus anguillicaudatus* individuals at LD2, FD1

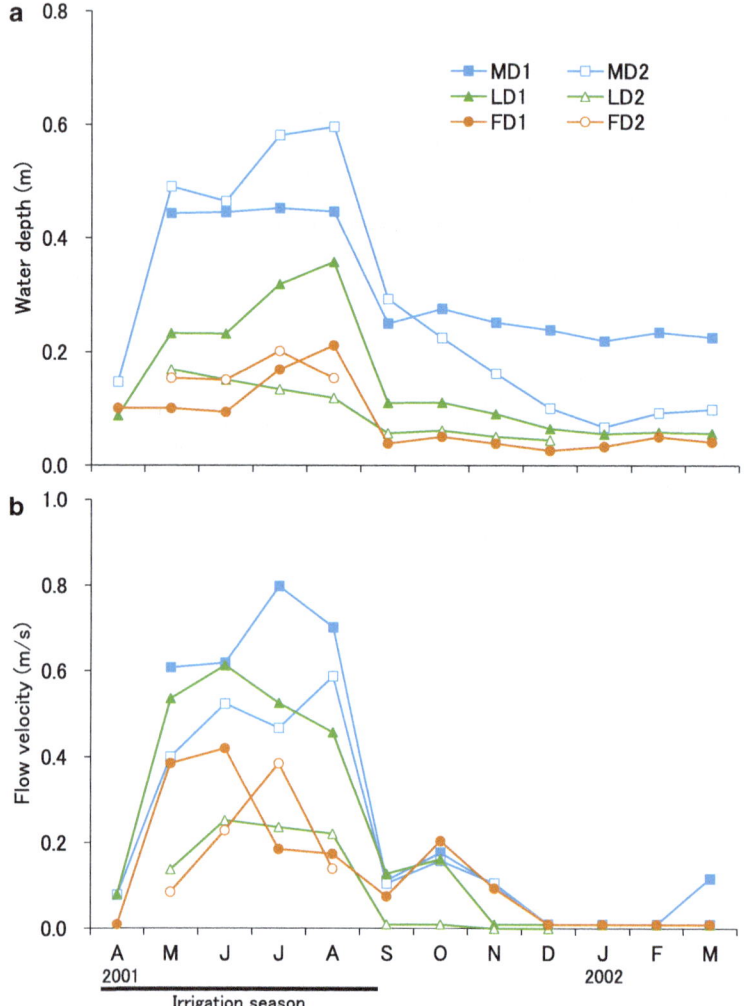

Fig. 10.9 Seasonal changes in **a** water depth and **b** flow velocity at the survey sites. *Source* Reprinted from Matsui and Satoh (2004a). Copyright 2004a

and FD2 were caught in months other than June in 2001 because of their abundance at that time.

Box plots and statistical results for fish biomass at the survey sites are shown in Fig. 10.12 and Table 10.3. Based on the Kruskal–Wallis test, *Opsariichthys platypus* and *Candidia temminckii* showed a significant difference. The Steel–Dwass test for *Opsariichthys platypus* and *Candidia temminckii* showed a significant difference between MD2 and FD1.

Illustrations of *Opsariichthys platypus* and *Misgurnus anguillicaudatus* are presented in Fig. 10.13.

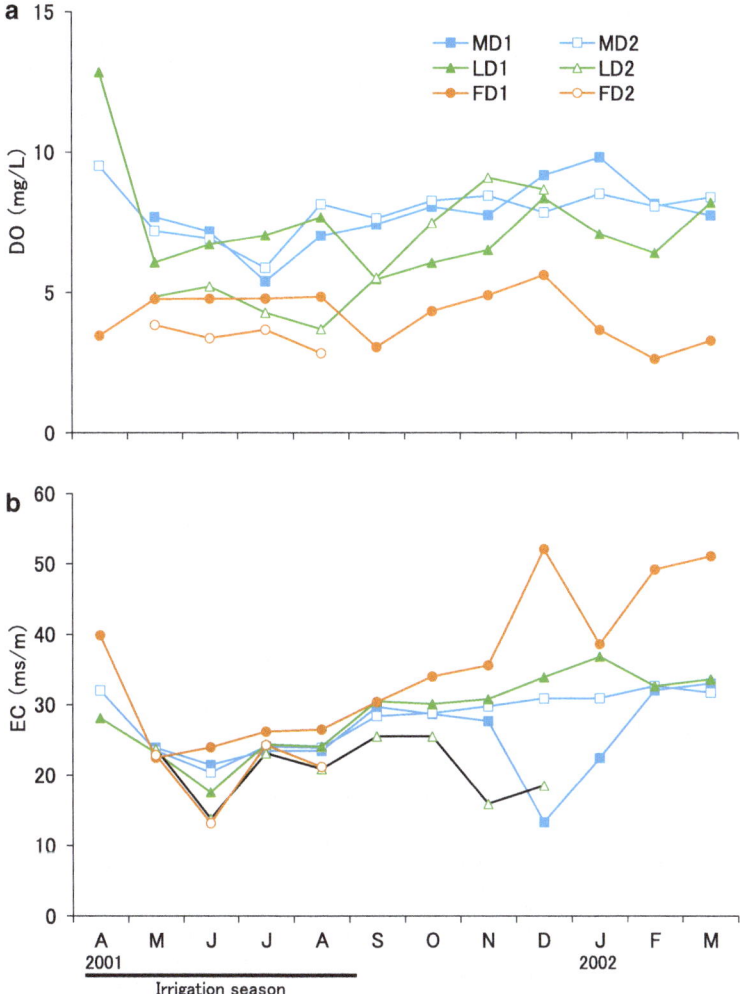

Fig. 10.10 Seasonal changes in **a** DO and **b** EC at the survey sites. *Source* Reprinted from Matsui and Satoh (2004a). Copyright 2004a

The information below shows the frequency distribution of the standard length of *Opsariichthys platypus* and *Misgurnus anguillicaudatus,* which were caught many times at the survey sites. Additionally, seasonal changes in the number and standard length (average ± standard deviation) and seasonal changes in the standard length distribution of *Opsariichthys platypus* and *Misgurnus anguillicaudatus* collected from the survey sites are also indicated.

Opsariichthys platypus

(1) **Frequency distribution of the standard length of *Opsariichthys platypus***

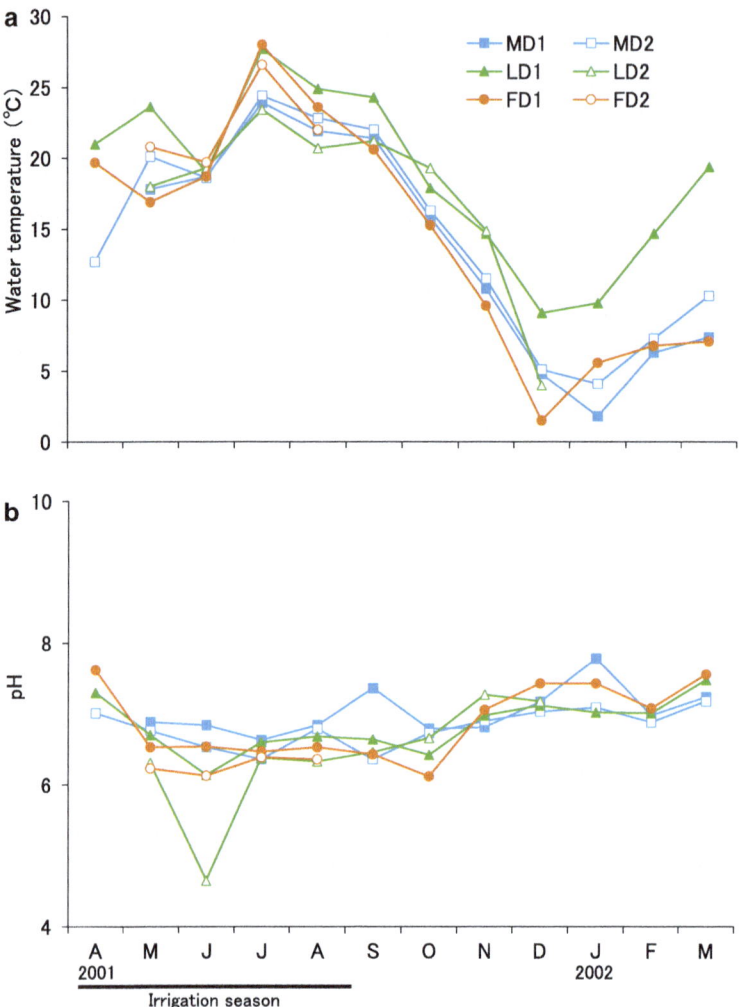

Fig. 10.11 Seasonal changes in **a** water temperature and **b** pH at the survey sites. *Source* Illustration based on Matsui and Satoh (2004a)

The frequency distribution of the standard length of *Opsariichthys platypus* at the survey sites is shown in Fig. 10.14. No individual was caught at FD1 and FD2. The fish harvest was large at MD2. The minimum and maximum values of the standard length of the individuals were 1.9 cm and 11.5 cm, respectively; thus, different standard lengths of the individuals were confirmed.

(2) **Seasonal changes in the standard length distribution of *Opsariichthys platypus***

Table 10.2 Total numbers of fish species caught at the survey sites

Species	MD1	MD2	LD1	LD2	FD1	FD2	Total
Cyprinus carpio	0	9	1	0	0	0	10
Carassius sp.	1	8	7	7	4	2	29
Opsariichthys platypus	16	140	10	4	0	0	170
Candidia temminckii	7	31	9	0	0	0	47
Tribolodon hakonensis	0	20	0	0	0	0	20
Pseudorasbora parva	0	0	1	0	0	0	1
Gnathopogon elongatus elongatus	8	8	20	4	5	3	48
Pseudogobio esocinus esocinus	0	2	0	0	0	0	2
Cyprinidae	21	9	0	5	1	0	36
Misgurnus anguillicaudatus	24	31	38	75	52	77	297
Silurus asotus	0	4	3	2	2	4	15
Oryzias latipes	0	1	0	0	0	0	1
Rhinogobius sp.	1	11	1	5	0	0	18
Total	78	274	90	102	64	86	694

Note All *Misgurnus anguillicaudatus* individuals at LD2, FD1 and FD2 were caught in months other than June in 2001 because of their abundance at that time
Source Reprinted from Matsui and Satoh (2004a). Copyright 2004a

The seasonal changes in the number and standard length (average ± standard deviation) of *Opsariichthys platypus* collected at each survey site are presented in Fig. 10.15. The number of individuals was high at MD2, but no individuals were found at FD1 and FD2. The average standard length was large from April to August and small after September at MD2.

The seasonal changes in the standard length distribution of *Opsariichthys platypus* collected from the survey sites are indicated in Fig. 10.16. There were 37 individuals harvested in April, but the number decreased to 5 individuals or less from May to August. In September, the number increased again to 34 individuals, and the standard length showed these were mainly small individuals. The standard length showed that only small individuals of approximately 3 cm were harvested after December until the following spring.

Misgurnus anguillicaudatus

(1) **Frequency distribution of the standard length of *Misgurnus anguillicau-datus***

The frequency distribution of the standard length of *Misgurnus anguilli-caudatus* at the survey sites is shown in Fig. 10.17. Individuals with standard lengths of 1–2 cm were caught at each survey site. In particular, the individuals whose standard length was more than 12 cm were harvested at LD2 and FD2.

(2) **Seasonal changes in the standard length distribution of *Misgurnus anguillicaudatus***

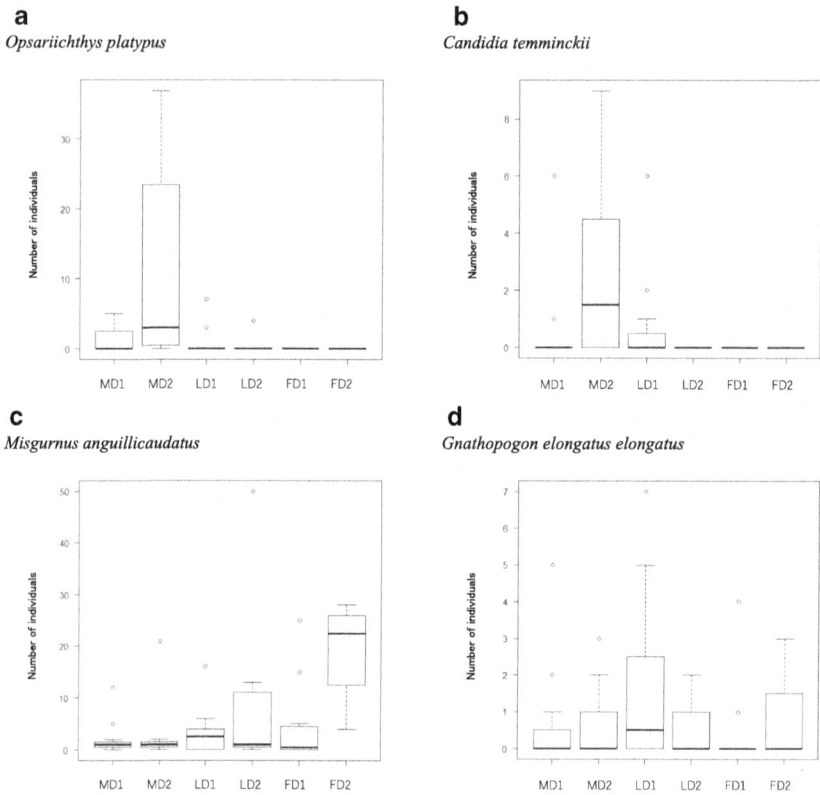

Fig. 10.12 Box plot of fish biomass at the survey sites. *Source* Illustration based on Matsui and Satoh (2004a)

The seasonal changes in the number and standard length (average ± standard deviation) of *Misgurnus anguillicaudatus* collected at each survey site are presented in Fig. 10.18. Individuals were found at LD1 and FD1 from May to October but not after November despite flowing water during the nonirrigation season. On the other hand, individuals were found at MD1 during the nonirrigation season.

The seasonal changes in the standard length distribution of *Misgurnus anguilli-caudatus* collected from the survey sites are indicated in Fig. 10.19. Five individuals whose standard lengths were less than 2 cm were caught in May, 152 individuals whose standard lengths were less than 6 cm were caught in June, and only several individuals were collected after September.

Table 10.3 Statistical results for fish biomass at the survey sites

a

Opsariichthys platypus
Kruskal–Wallis test P value = 0.000
Steel–Dwass test

	MD1	MD2	LD1	LD2	FD1	FD2
MD1		0.458	0.813	0.725	0.107	0.627
MD2			0.062	0.107	*0.004*	0.201
LD1				0.999	0.699	0.959
LD2					0.824	0.981
FD1						NaN
FD2						

b

Candidia temminckii
Kruskal–Wallis test P value = 0.004
Steel–Dwass test

	MD1	MD2	LD1	LD2	FD1	FD2
MD1		0.366	0.998	0.817	0.657	0.950
MD2			0.482	0.114	*0.030*	0.420
LD1				0.671	0.460	0.894
LD2					NaN	NaN
FD1						NaN
FD2						

c

Misgurnus anguillicaudatus
Kruskal–Wallis test P value = 0.112
Steel–Dwass test

	MD1	MD2	LD1	LD2	FD1	FD2
MD1		0.999	0.997	0.987	0.999	0.081
MD2			0.983	0.992	0.999	0.055
LD1				0.997	0.999	0.098
LD2					0.974	0.522
FD1						0.191
FD2						

d

Gnathopogon elongatus elongatus
Kruskal–Wallis test P value = 0.559
Steel–Dwass test

	MD1	MD2	LD1	LD2	FD1	FD2
MD1		0.995	0.841	0.999	0.988	1.000
MD2			0.950	0.999	0.841	0.999

(continued)

Table 10.3   (continued)

d

Gnathopogon elongatus elongatus
Kruskal–Wallis test P value = 0.559
Steel–Dwass test

LD1				0.912	0.494	0.977
LD2					0.942	0.999
FD1						0.999
FD2						

Note Italics indicates a significant difference
Source Illustration based on Matsui and Satoh (2004a)

a *Opsariichthys platypus*

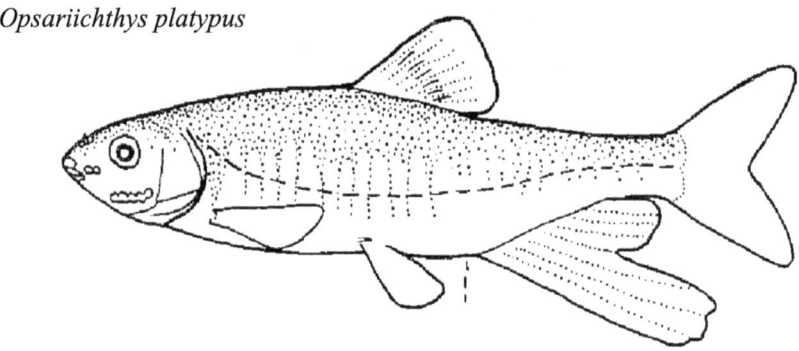

b *Misgurnus anguillicaudatus*

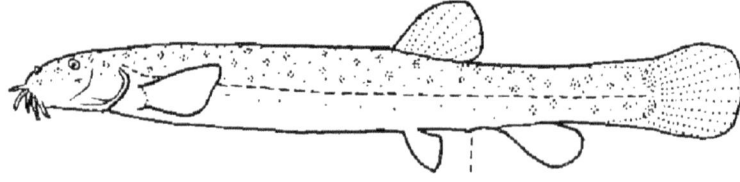

Fig. 10.13   Illustrations of the fishes caught at the survey sites. *Note* The standard length of **a** *Opsariichthys platypus* and **b** *Misgurnus anguillicaudatus* was 13 cm and 10 cm, respectively. *Source* Reprinted from Nakabo (2013). Copyright 2013 Tokai University Publishing Division

10.3.3   Total Numbers of Aquatic Insects Caught at the Survey Sites

In total, 792 individual aquatic insects were caught at all survey sites. The species were as follows: *Ischnura* sp., *Calopteryx atlata*, *Sieboldius albardae*, *Anisogomphus maackii*, *Anax partenope julius*, *Macromia amphigena*, *Orthetrum albistyrum*

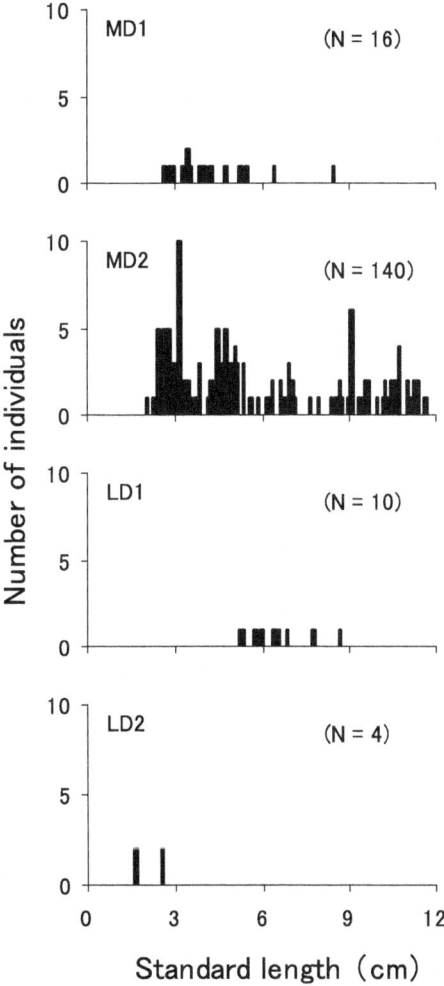

Fig. 10.14 Frequency distribution of the standard length of *Opsariichthys platypus* at the survey sites. *Note N* is the total number of individuals caught at each site. *Source* Reprinted from Matsui and Satoh (2004a). Copyright 2004a

speciosum, *Sympetrum infuscatum*, *Sympetrum baccha matutinum*, *Laccotrephes japonensis* and *Ranatra chinensis* (Table 10.4). The values in the table show the actual number of aquatic insects caught at the survey sites. The order of listed aquatic insects follows Kawai (1985).

Box plots and statistical results for dragonfly biomass at the survey sites are shown in Fig. 10.20 and Table 10.5. As a result of the Kruskal–Wallis test, *Calopteryx atlata*, *Orthetrum albistyrum speciosum* and *Sympetrum infuscatum* showed significant differences. The Steel–Dwass test for *Calopteryx atlata* showed a significant difference between FD1 and MD2 and FD1 and LD1. In addition, the Steel–Dwass test for *Orthetrum albistyrum speciosum* showed a significant difference between FD1 and MD1, FD1 and MD2 and FD1 and LD2, that for *Sympetrum infuscatum*

Fig. 10.15 Seasonal
changes in the number and
standard length (average ±
standard deviation) of
Opsariichthys platypus
collected at each survey site.
Notes Vertical lines show the
standard deviation. *N* is the
number of individuals
collected at each site.
*indicates no survey. *Source*
Reprinted from Matsui
(2009). Copyright 2009

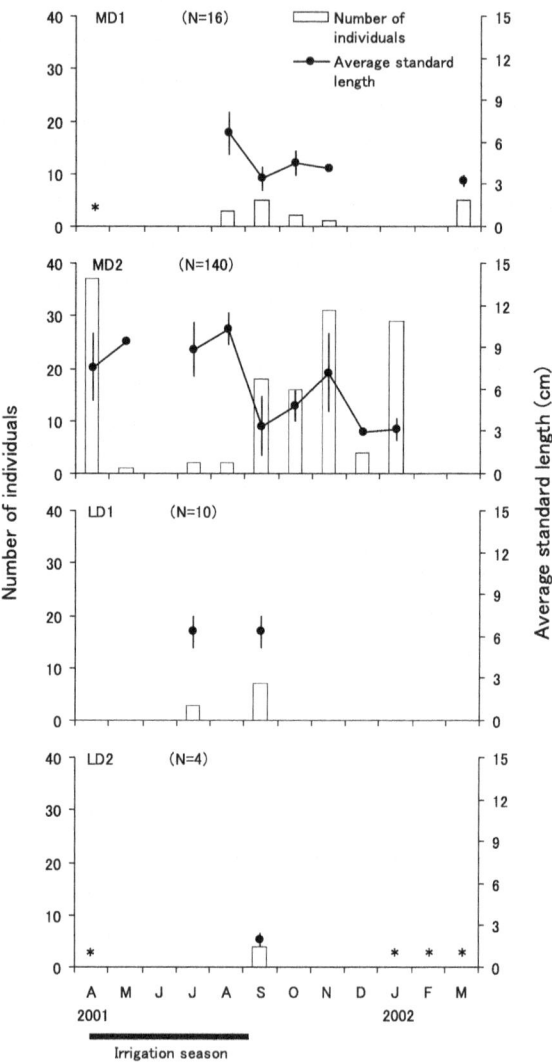

showed a significant difference between FD2 and MD1, FD2 and MD2 and FD2 and
LD1.

The illustrations of *Calopteryx atlata* and *Orthetrum albistyrum speciosum* are
presented in Fig. 10.21.

The information below shows the frequency distribution of the head width of
Calopteryx atrata and *Orthetrum albistylum speciosum,* which were caught many
times at the survey sites. Additionally, seasonal changes in the number, head width
(average ± standard deviation) and head width distribution of *Calopteryx atrata* and
Orthetrum albistylum speciosum collected from the survey sites are also indicated.

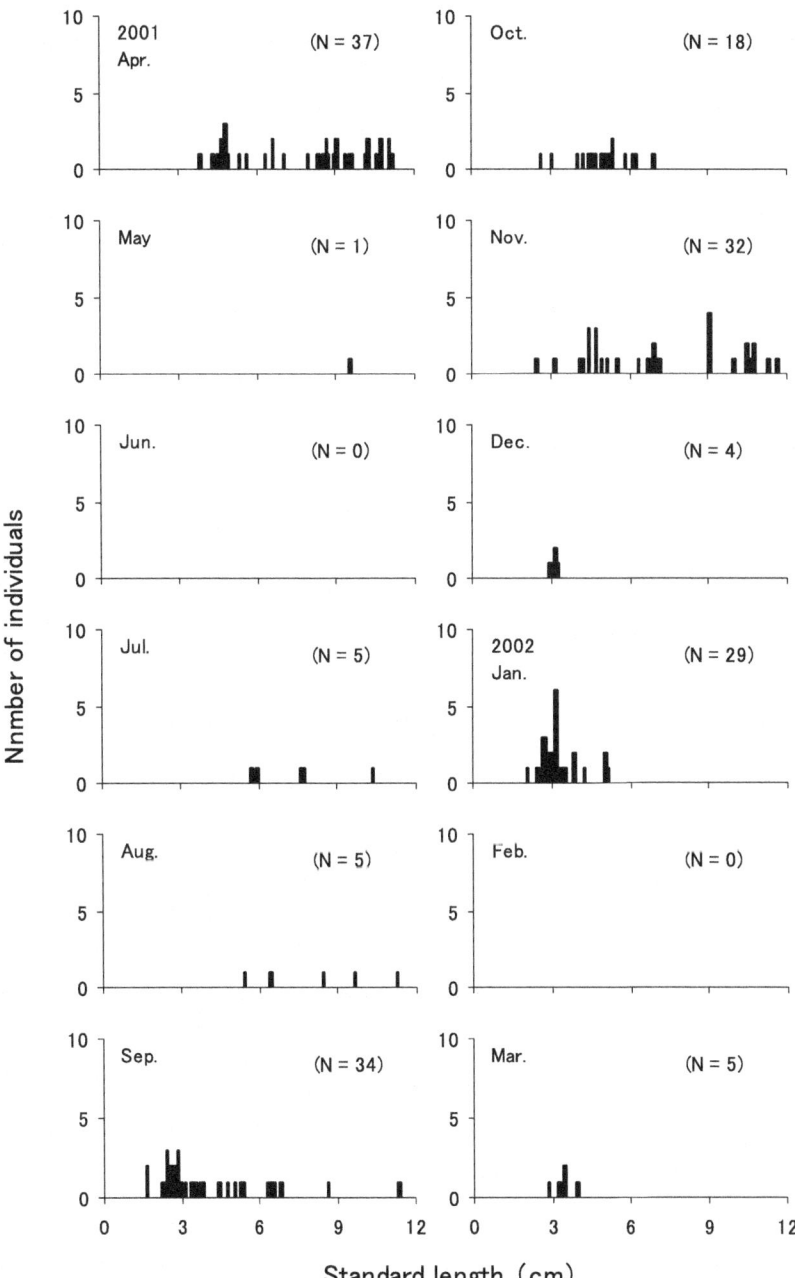

Fig. 10.16 Seasonal changes in the standard length distribution of *Opsariichthys platypus* collected from the survey sites. *Note N* is the total number of individuals collected each month. *Source* Reprinted from Matsui (2009). Copyright 2009

Fig. 10.17 Frequency
distribution of the standard
length of *Misgurnus
anguillicaudatus* at the
survey sites. *Notes* All
Misgurnus anguillicaudatus
individuals at LD2, FD1 and
FD2 were caught in months
other than June in 2001
because of their abundance
at that time. N is the total
number of individuals caught
at each site. *Source*
Reprinted from Matsui and
Satoh (2004a). Copyright
2004a

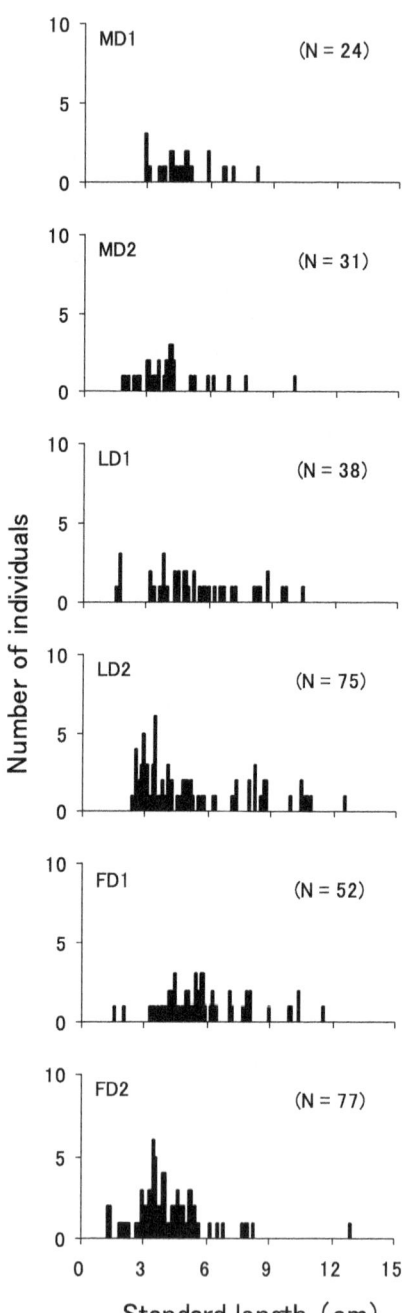

Fig. 10.18 Seasonal changes in the number and standard length (average ± standard deviation) of *Misgurnus anguillicaudatus* collected at each survey site. *Notes* Vertical lines show the standard deviation. *N* is the number of individuals collected at each site. *indicates no survey. *Source* Reprinted from Matsui (2009). Copyright 2009

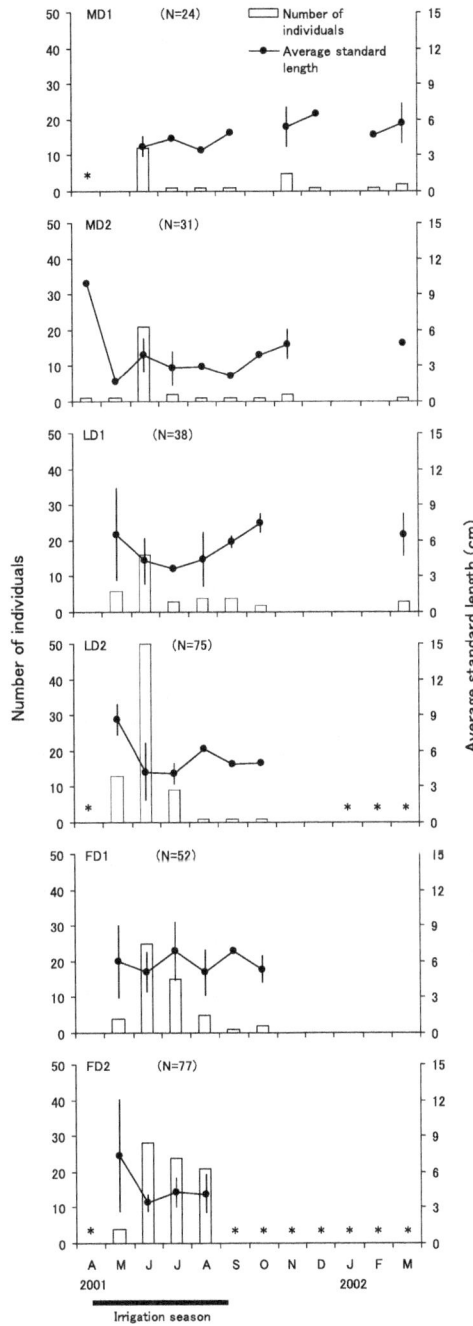

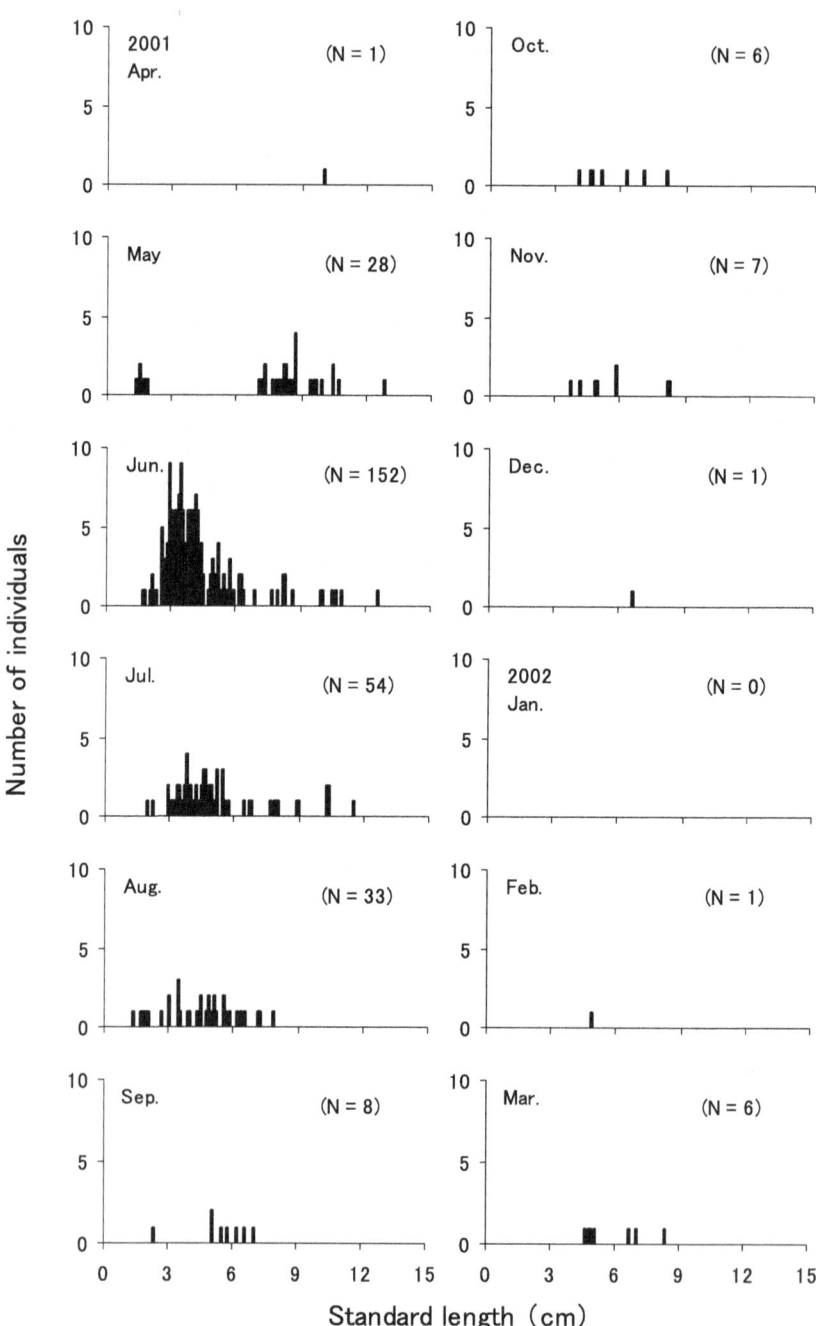

Fig. 10.19 Seasonal change in the standard length distribution of *Misgurnus anguillicaudatus* collected from the survey sites. *Note N* is the total number of individuals collected each month. *Source* Reprinted from Matsui (2009). Copyright 2009

Table 10.4   Total numbers of aquatic insects caught at the survey sites

Species	MD1	MD2	LD1	LD2	FD1	FD2	Total
Ischnura sp.	1	0	9	3	1	3	17
Coenagrionidae	0	1	1	2	0	0	4
Calopteryx atlata	26	71	266	12	1	0	376
Sieboldius albardae	1	0	0	0	0	0	1
Anisogomphus maackii	5	10	1	0	0	0	16
Anax partenope julius	0	1	3	0	0	0	4
Macromia amphigena	2	3	3	0	0	0	8
Orthetrum albistyrum speciosum	3	10	64	8	238	0	323
Sympetrum infuscatum	0	0	0	1	5	26	32
Sympetrum baccha matutinum	0	0	0	0	3	0	3
Laccotrephes japonensis	2	0	2	0	0	2	6
Ranatra chinensis	1	0	1	0	0	0	2
Total	41	96	350	26	248	31	792

Source Reprinted from Matsui and Satoh (2004a). Copyright 2004a

Calopteryx atrata

(1)   Frequency distribution of the head width of *Calopteryx atrata*

The frequency distribution of the head width of *Calopteryx atrata* at the survey sites is shown in Fig. 10.22. No individual was caught at FD2. The aquatic insect harvest was large at LD1. The head width size of the individuals caused some peaks at approximately 1–2 mm, 2–3 mm and 4 mm; thus, various head width sizes of the individuals were confirmed.

(2)   Seasonal changes in the head width of *Calopteryx atrata*

The seasonal changes in the number and head width (average ± standard deviation) of *Calopteryx atlata* collected at each survey site are presented in Fig. 10.23. The number of individuals was high at LD1, but individuals were low at FD2. The average head width was large from May to June and small from October to March at MD2 and LD1.

The seasonal changes in the head width distribution of *Calopteryx atlata* collected from the survey sites are indicated in Fig. 10.24. The relationship between head width and wet weight of *Calopteryx atlata* for each group is shown in Table 10.6. The head width distribution was classified into three types, in descending order by the A1, A2 and A3 groups.

Orthetrum albistylum speciosum

(1)   Frequency distribution of the head width of *Orthetrum albistylum speciosum*

The frequency distribution of the head width of *Orthetrum albistylum speciosum* at the survey sites is shown in Fig. 10.25. The aquatic insect harvest

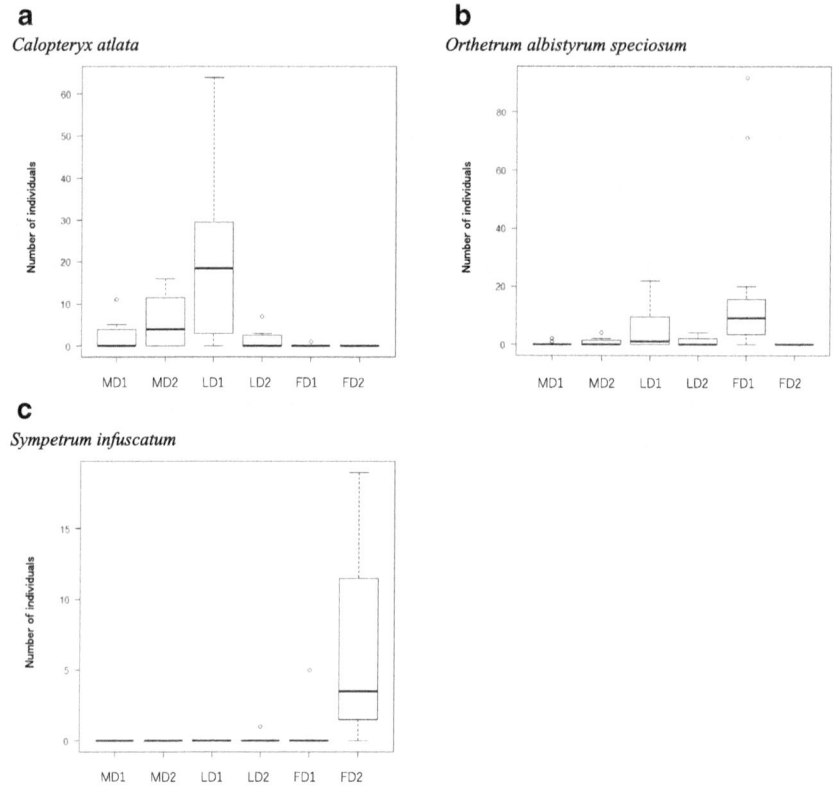

Fig. 10.20 Box plot of dragonfly biomass at the survey sites. *Source* Illustration based on Matsui and Satoh (2004a)

was large at FD1. No individual was caught at FD2. The head width size of the individuals caused some peaks at 1–5 mm; thus, various head width sizes of the individuals were confirmed.

(2)   **Seasonal changes in the head width of *Orthetrum albistylum speciosum***

The seasonal changes in the number and head width (average ± standard deviation) of *Orthetrum albistyrum speciosum* collected at each site are presented in Fig. 10.26. The number of individuals was high at FD1, but no individuals were found at FD2. The average head width was large from April to July and small after September until the next spring at FD1.

The seasonal changes in the head width distribution of *Orthetrum albistyrum speciosum* collected from the survey sites are indicated in Fig. 10.27. The relationship between head width and wet weight of *Orthetrum albistyrum speciosum* for each group is shown in Table 10.7. Head width distribution was classified into five types, in descending order by the B1, B2, B3, B4 and B5 groups.

Table 10.5 Statistical results for dragonfly biomass at the survey sites

a
Calopteryx atlata
Kruskal–Wallis test P value = 0.000
Steel–Dwass test

	MD1	MD2	LD1	LD2	FD1	FD2
MD1		0.717	0.096	0.997	0.260	0.627
MD2			0.419	0.558	*0.027*	0.301
LD1				0.144	*0.009*	0.201
LD2					0.524	0.764
FD1						0.992
FD2						

b
Orthetrum albistyrum speciosum
Kruskal–Wallis test P value = 0.000
Steel–Dwass test

	MD1	MD2	LD1	LD2	FD1	FD2
MD1		0.812	0.323	0.894	*0.002*	0.950
MD2			0.887	1.000	*0.005*	0.678
LD1				0.890	0.353	0.416
LD2					*0.031*	0.764
FD1						0.072
FD2						

c
Sympetrum infuscatum
Kruskal–Wallis test P value = 0.000
Steel–Dwass test

	MD1	MD2	LD1	LD2	FD1	FD2
MD1		NaN	NaN	0.850	0.931	*0.025*
MD2			NaN	0.824	0.918	*0.017*
LD1				0.824	0.918	*0.017*
LD2					0.999	0.184
FD1						0.133
FD2						

Note Italics indicates a significant difference
Source Illustration based on Matsui and Satoh (2004a)

a *Calopteryx atlata*

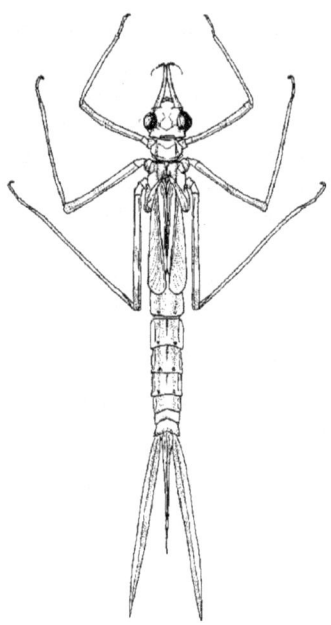

b *Orthetrum albistyrum speciosum*

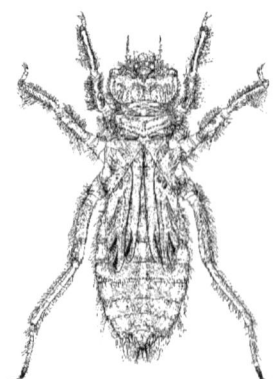

Fig. 10.21 Illustrations of aquatic insects caught at the survey sites. *Note* For head width and wet weight of **a** *Calopteryx atlata* and **b** *Orthetrum albistyrum speciosum* for each group collected from the survey sites, see Tables 10.6 and 10.7, respectively. *Source* Reprinted from Kawai (1985). Copyright 1985 Tokai University Publishing Division; Kawai and Tanida (2005). Copyright 2005 Tokai University Publishing Division

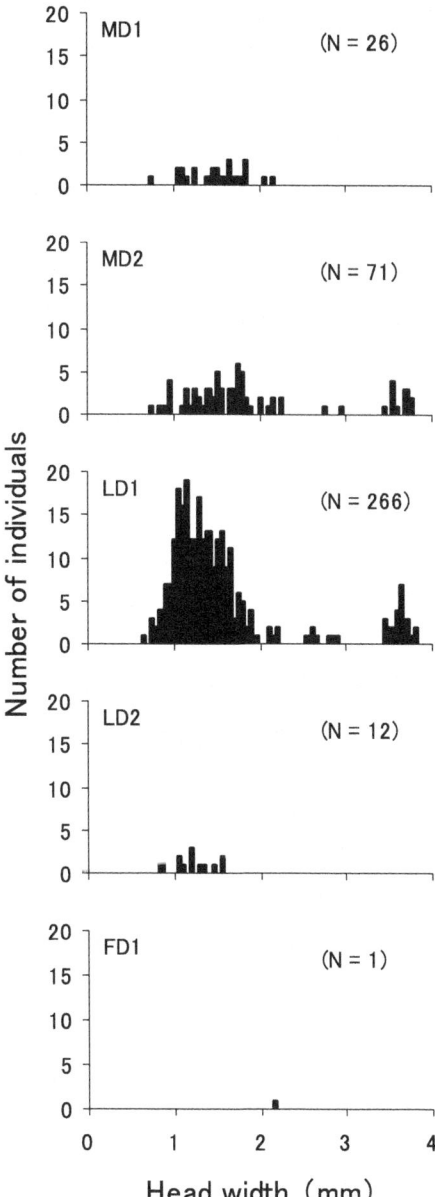

Fig. 10.22 Frequency distribution of the head width of *Calopteryx atrata* at the survey sites. *Note N* is the total number of individuals caught at each site. *Source* Reprinted from Matsui and Satoh (2004a). Copyright 2004a

Fig. 10.23 Seasonal changes in the number and head width (average ± standard deviation) of *Calopteryx atlata* collected at each survey site. *Notes* Vertical lines show the standard deviation. *N* is the number of individuals collected at each site. *indicates no survey. *Source* Reprinted from Matsui (2009). Copyright 2009

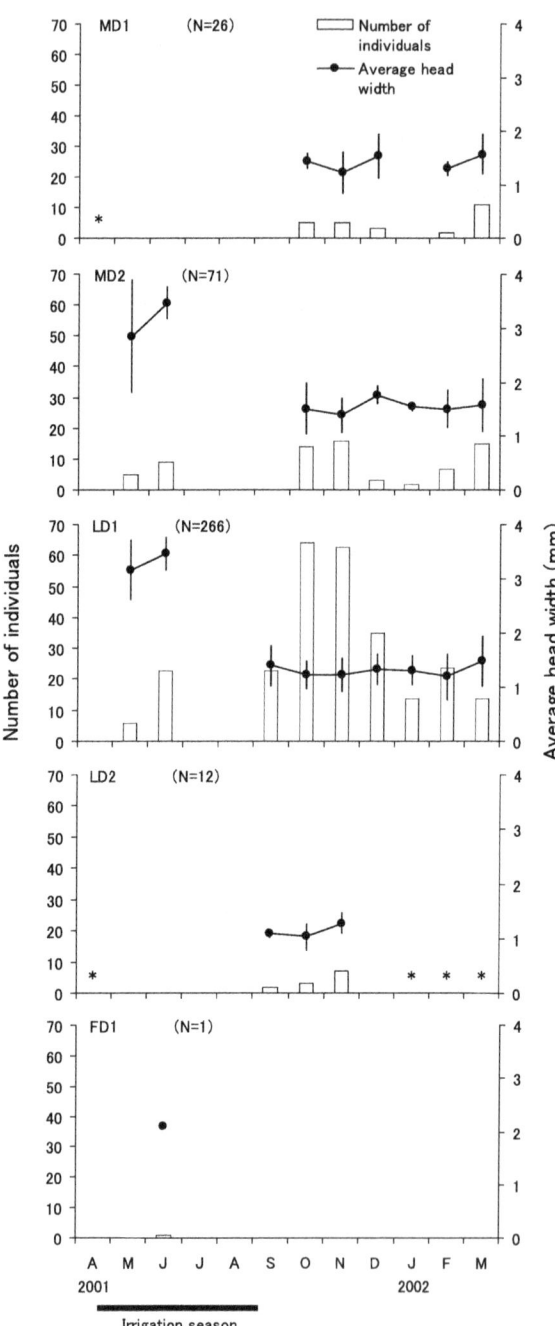

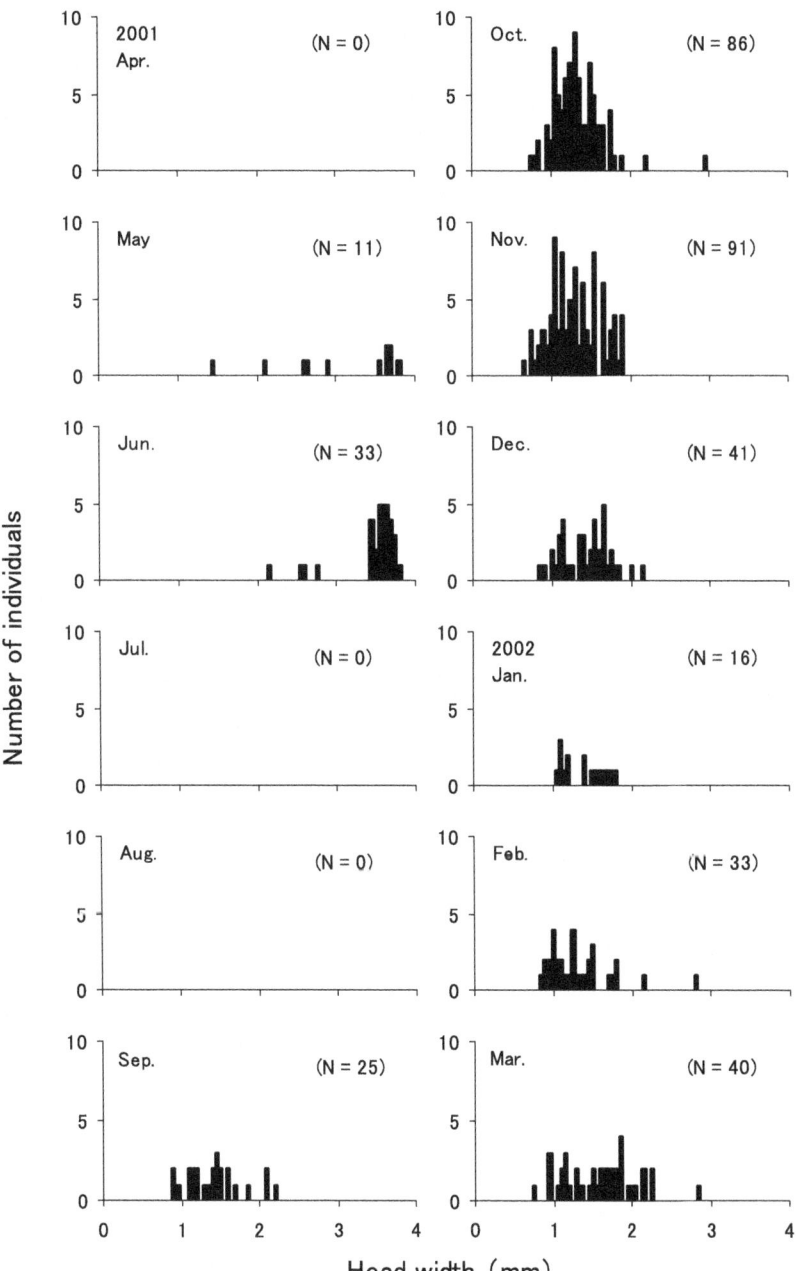

Fig. 10.24 Seasonal changes in the head width distribution of *Calopteryx atlata* collected from the survey sites. *Note N* is the total number of individuals collected in each month. *Source* Reprinted from Matsui (2009). Copyright 2009

Table 10.6 Head width and wet weight of *Calopteryx atlata* for each group

Group	Number of individuals	Head width		Wet weight	
		Range (mm)	Average ± SD (mm)	Range (g)	Average ± SD (g)
A1	35	3.40–3.75	3.57 ± 0.10	0.106–0.188	0.150 ± 0.022
A2	9	2.50–2.90	2.69 ± 0.15	0.052–0.090	0.069 ± 0.013
A3	332	0.60–2.20	1.32 ± 0.33	0.001–0.033	0.006 ± 0.005

Source Reprinted from Matsui (2009). Copyright 2009

10.4 Discussion

10.4.1 Physical Environment in the Drainage Canal System

The seasonal changes in water depth and flow velocity demonstrated the features of drainage water for irrigation season from the middle of April to the beginning of September. As the water depth at MD1 was greater, the flow velocity at MD1 was lower than that at the other sites during the nonirrigation season. Thus, the lower reaches of the survey area were estimated to have been in a flooded state.

In comparison with the other sites, at FD1, the DO significantly decreased starting in September, and the EC remarkably increased starting in October; thus, FD1 was found to be polluted. The gray water flowed into FD1 and was not diluted because there was no drainage from the paddy fields during the nonirrigation season. Therefore, it was assumed that water quality deterioration had occurred since September.

There was an approximately 0.30 m difference in elevation at a joint of lateral drain and farm drain; however, no issues existed related to movement of the fishes. Because the water depth at LD1 exceeded 0.30 m with precipitation, the surface of the water between LD1 and FD1 and between LD2 and FD2 was connected. As described above information, the water depth, flow velocity and water quality of the drains were significantly different between the irrigation season and the nonirrigation season in this survey area.

10.4.2 Ecology of Opsariichthys Platypus

(1) **Habitat distribution**

In this study, 170 individuals of *Opsariichthys platypus* were caught at all 6 survey sites, and of these, 140 individuals, or approximately 82%, were harvested at MD2. On the other hand, 16 individuals were caught at MD1, which was a three-sided, concrete-lined waterway in the same main drain.

Fig. 10.25 Frequency
distribution of the head width
of *Orthetrum albistylum
speciosum* at the survey sites.
Note N is the total number of
individuals caught at each
site. *Source* Reprinted from
Matsui and Satoh (2004a).
Copyright 2004a

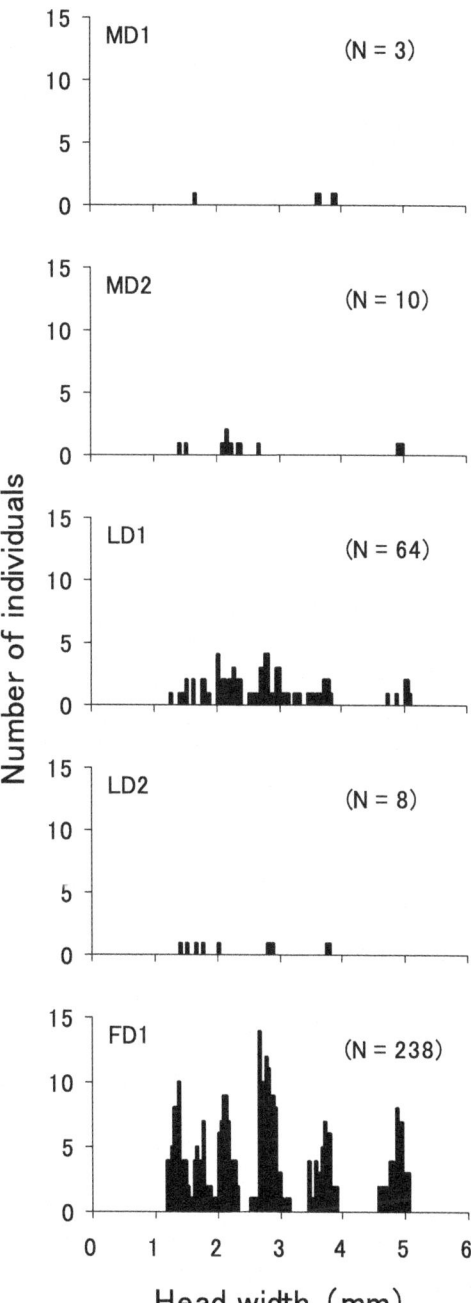

Fig. 10.26 Seasonal
changes in the number and
head width (average ±
standard deviation) of
*Orthetrum albistyrum
speciosum* collected at each
survey site. *Notes* Vertical
lines show the standard
deviation. N is the number of
individuals collected at each
site. *indicates no survey.
Source Reprinted from
Matsui (2009). Copyright
2009

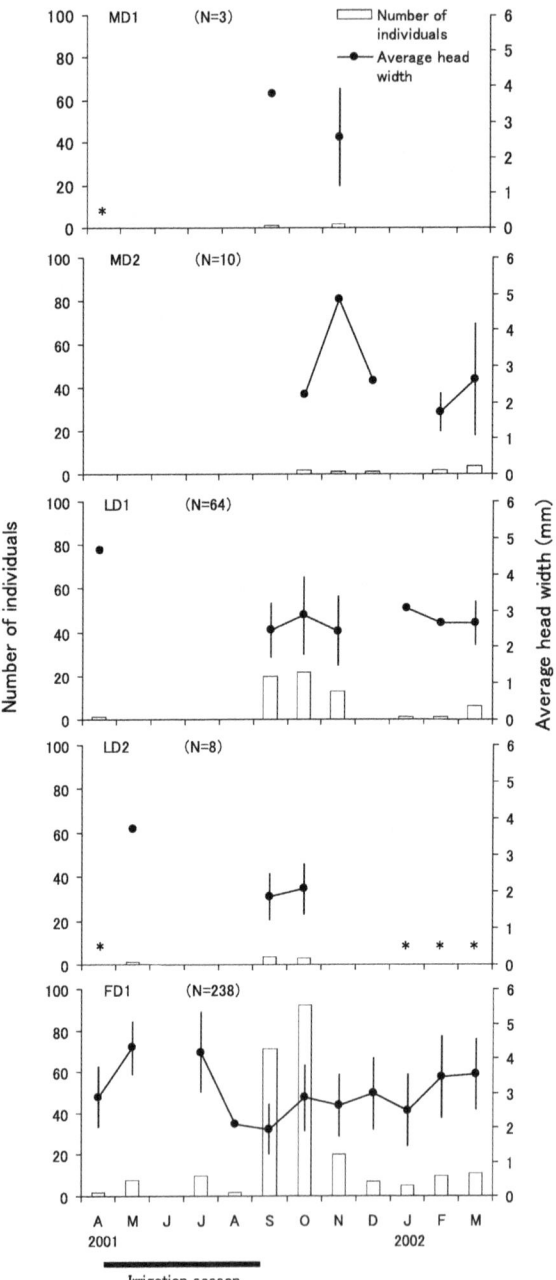

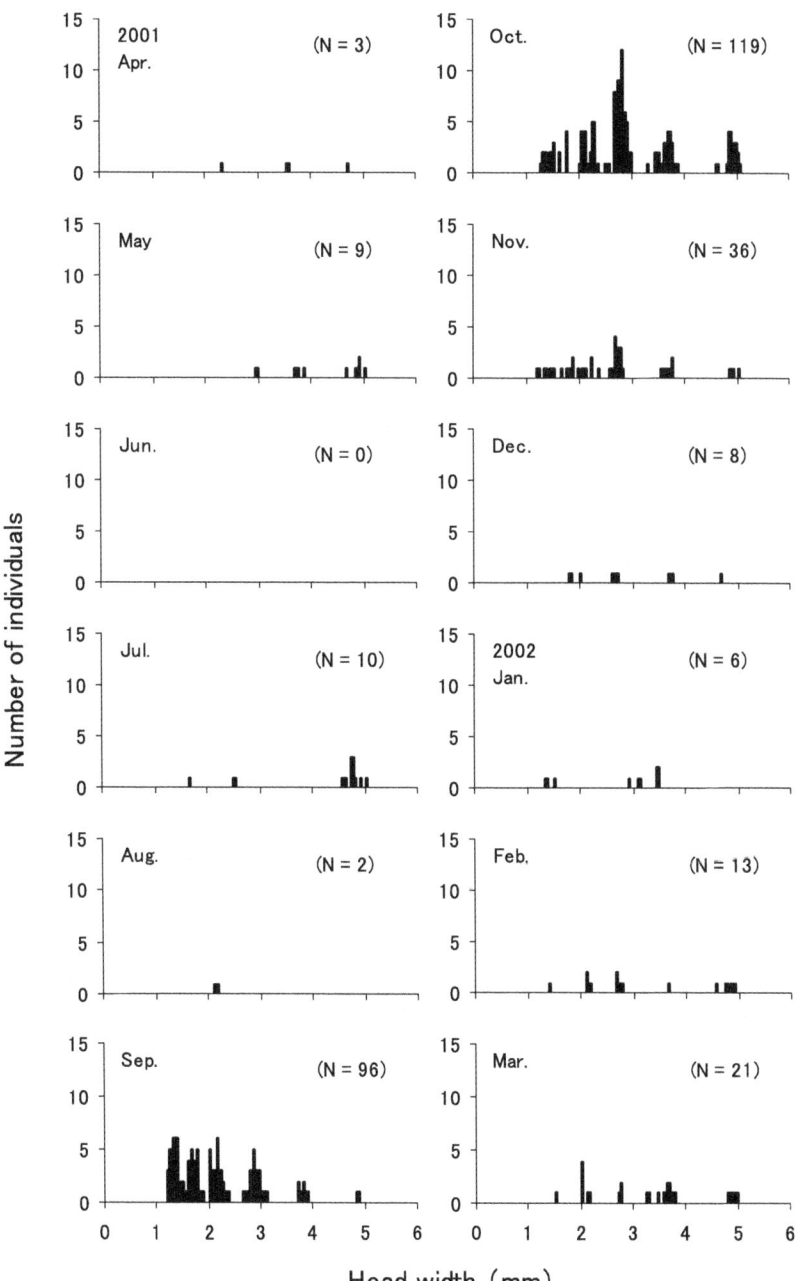

Fig. 10.27 Seasonal changes in the head width distribution of *Orthetrum albistyrum speciosum* collected from the survey sites. *Note N* is the total number of individuals collected in each month. *Source* Reprinted from Matsui (2009). Copyright 2009

Table 10.7 Head width and wet weight of *Orthetrum albistyrum speciosum* for each group

Group	Number of individuals	Head width		Wet weight	
		Range (mm)	Average ± SD (mm)	Range (g)	Average ± SD (g)
B1	41	4.50–5.00	4.79 ± 0.13	0.234–0.659	0.377 ± 0.086
B2	47	3.20–3.80	3.59 ± 0.14	0.118–0.286	0.179 ± 0.036
B3	97	2.45–3.05	2.72 ± 0.12	0.039–0.135	0.074 ± 0.018
B4	61	1.90–2.30	2.08 ± 0.10	0.017–0.059	0.032 ± 0.008
B5	77	1.15–1.80	1.46 ± 0.20	0.002–0.023	0.010 ± 0.005

Source Reprinted from Matsui (2009). Copyright 2009

Additionally, the 140 individuals caught at MD2 had a variety of standard length sizes.

As this species lays its eggs on the gravel and sand bottoms of riffles, where the water depth was 5–10 cm and the water flow loosened (Miyaji et al. 1963; Mori and Nagoshi 1989), it is likely that this species mainly inhabits MD2 in the drainage system. It is an important condition that bottom material is gravel and sand to ensure spawning grounds exist for this species. Therefore, this factor should be included when implementing paddy farmland consolidation in the future.

(2) **Life history**

Regarding the seasonal changes in the standard length distribution of this species, yearling fish (0-year-old fish) are 7.8–9 cm until autumn, 1-year-old fish are 9.8–11.3 cm, 2-year-old fish are 11.3–12.4 cm and 3-year-old fish are 12.5–14.0 cm (Nakamura 1952). As 34 individuals were harvested in September in this study, based on the above information, it is estimated that the one individual whose standard maximum length was 11.2 cm was a 1-year-old fish, and the remaining fish were 0 years old.

It was determined that 0-year-old fish appeared in September for the first time in this area. Given that the spawning season of this species is said to be from May to August (Miyaji et al. 1963; Mori and Nagoshi 1989), this result is lagging.

In terms of the downstream side of this district, it is difficult for this species to ascend from the Gogyo River to the main drain because the difference in elevation is large; thus, the upstream water-level constant control gate is dropped for the irrigation season.

For the upstream side, it is conceivable that the fish living in the Gogyo River move into the main canal from the Akaido diversion weir through the paddy fields and then to the main drain. However, considering that this species was less abundant in the lateral drain and were not harvested in the farm drain, the supply of this fish from Gogyo River is likely to be small. Thus, it was estimated that this species mainly inhabits and grows in the drainage canal system with their entire and continuous processes of growth, including spawning, hatching, juvenile and adult fishes, occurring there, especially at MD2.

(3) **Habitat in the nonirrigation season**

In this study, the 1- to 3-year-old fish were caught in winter (Mizuno and Nakagawa 1968), while they were rarely harvested, except for yearling fish (0-year-old fish), in the nonirrigation season. In addition, the 0-year-old fish caught in the area around MD2 showed a tendency to be collected even at MD1. Mori and Nagoshi (1989) noted that immature and adult fishes like riffles in rivers and move to deep or hydrophytic areas in winter. Judging from the growth process reported by Okuda et al. (1996), the yearling fish (0-year-old fish) that were born in the vicinity of MD2 in September were estimated to be immature fish in October.

Regarding the physical environment of the drainage canal system, in comparison with in the irrigation season, in the nonirrigation season after October, the water depth and flow velocity were significantly reduced. In particular, the water depth around MD1 was maintained at approximately 30 cm for the nonirrigation season, while those at other survey sites showed a significantly smaller value. Thus, it is believed that this species moved to the downstream portion where the water depth was high in the nonirrigation season.

For example, for the medaka fish *Oryzias latipes*, the number of individuals did not decrease in the concrete-lined ditches of consolidated paddy fields when they did not experience a water-level drop during the nonirrigation season after autumn. On the other hand, the number of individuals dramatically declined in the unlined ditches of nonconsolidated paddy fields where the water level was lowered. Thus, the fish escaped to the main canal due to the water-level drop, or those that moved to the remaining water experienced a high mortality rate (Tsujii and Ueda 2003). Even for *Opsariichthys platypus* in this study, it is conceivable that they moved to the deep and aquatic vegetation of the nonconsolidated paddy fields located in the Gogyo River confluence area, approximately 3 km downstream from MD1.

10.4.3 Ecology of Misgurnus Anguillicaudatus

(1) **Habitat distribution**

In this study, 297 individuals of *Misgurnus anguillicaudatus* were harvested at all 6 survey sites, and the number of individuals caught at the low-water depth sites FD1, FD2 and LD2 was high. Additionally, adult fish whose standard lengths were more than 12 cm were caught at FD2 and LD2.

As this species lays its eggs on aquatic plants in the minor groove and rice plants in paddy fields (Miyaji et al. 1963), it was estimated that this species mainly inhabited the lateral and farm drains in the drainage canal system. Saitoh et al. (1988) noted that this species invades and spawns in temporary water areas. Because FD2 and LD2 are temporary water areas, these survey results are consistent with the abovementioned results.

To ensure spawning grounds for this species exist, it is an important that the elevation difference between the main, lateral and farm drains or farm drains and paddy fields is small. Thus, the information described above should be considered when implementing paddy farmland consolidation in the future. In addition, it is specifically important to ensure continuity of the farm drains to the paddy fields to expand the shallow water area, which is a suitable condition for spawning. In addition, shallow farm drains can be effectively constructed by separating surface drainage and underground drainage as proposed by Shinzawa and Koide (1963). It is also desirable to reevaluate these efforts from the ecological conservation point of view. Specific engineering countermeasures will be described later (Fig. 11.15).

(2)   **Life history**

Regarding the seasonal changes in the standard length distribution of this species, the standard length reached 2 cm at 10 days, 8–10 cm at 1 year and 10–12 cm at 2 years (Miyaji et al. 1963). As the spawning season of this species was from April to June (Miyaji et al. 1963), the 5 individuals caught in May that were not 2 cm were estimated to be yearling fish (0-year-old fish). On the other hand, the individuals whose standard length was 3–6 cm were found in larger numbers in June, and they were not found in May because the yearling fish (0-year-old fish), which were born in May, rapidly grew in June.

In May, the yearling fish (0-year-old fish) appeared for the first time in this area. It was estimated that this species was spawning and hatching, especially around the lateral and farm drains in the drainage canal system. Thus, the drainage canal system was believed to function as a spawning ground and habitat for this species.

(3)   **Habitat in the nonirrigation season**

This species moved upstream to spawn in the rainy season and downstream to overwinter in autumn (Kubota 1961). In this study, the number of individuals significantly increased at the LD2, FD1 and FD2 sites located upstream in the irrigation season, while this species was not caught during the nonirrigation season. Despite having year-round water flow at FD1, no individuals of this species were collected from November until the following spring.

On the other hand, the number of individuals caught at MD1 and MD2, which are downstream, in the nonirrigation season was significantly less than that caught at LD2, FD1 and FD2, which are upstream, in the irrigation season. Thus, it is likely that they died at those sites. More specifically, the main drain of the drainage canal system was concrete on three sides, whereas the lateral and farm drains were concrete on two sides (canal bed was natural); thus, the fish likely overwintered in the lateral and farm drains by diving in mud. In fact, after digging into a canal bed at FD2 that had dried in the nonirrigation season, a few overwintering individuals were found. From the above information, this species was determined to overwinter in mud in the farm drain or descend to the downstream portion rather than overwinter at MD1 in the nonirrigation season, which is similar to the scenario for *Opsariichthys platypus*.

10.4.4 Ecology of Calopteryx Atlata

(1) **Habitat distribution**

In this study, 376 individuals of *Calopteryx atlata* were caught at all 6 survey sites, and 266 individuals, or approximately 71%, were harvested at LD1. On the other hand, 12 individuals were caught only at LD2, which had no water flow from January to April in the nonirrigation season in the same lateral drain. Additionally, the 266 individuals caught at LD1 had a variety of head width sizes.

As an ecological feature of this species, its habitat is widely distributed from the middle basin in the river to the lower basin of the plains (Tsuda and Rokuyama 1973). Additionally, this species lives in areas with gently flowing water where aquatic plants flourish from the plains to hilly areas, and it prefers running water areas (Ishida and Ishida 1985). Thus, this species likely inhabits LD1, where there is running water in the nonirrigation season in the lateral drain consistent with its need for a flowing water environment. Running water in the nonirrigation season is an important condition for the survival of this species. Therefore, this information should be considered when implementing farmland consolidation in the future.

(2) **Life history**

Considering the seasonal changes in the head width distribution of larvae, this species is estimated to emerge in June. In particular, this species will spawn and hatch near LD1 in the drainage canal system. Thus, the drainage canal system is believed to function as spawning ground and habitat for this species.

10.4.5 Ecology of Orthetrum Albistyrum Speciosum

(1) **Habitat distribution**

In this study, 323 individuals of *Orthetrum albistyrum speciosum* were caught at all 6 survey sites, and 238 individuals, or approximately 74%, were harvested at FD1. On the other hand, no individual was caught at FD2, which had no water flow from September to April in the nonirrigation season. Additionally, the 238 individuals caught at FD1 had a variety of head width sizes.

As an ecological feature of this species, this species inhabits ponds (Tsuda and Rokuyama 1973); additionally, it inhabits puddles, paddy fields, ditches, etc., from plains to low mountains, and it prefers areas with flowing water (Ishida and Ishida 1985). Thus, it is estimated that this species lives at FD1, where there is flowing water in the nonirrigation season in the farm drain corresponding to its preferred environment. The significance of flowing water in the nonirrigation season is the same as that for *Calopteryx atlata*.

(2)   **Life history**

Based on the seasonal changes in the head width distribution of larvae, there are two generations in a year: an overwintering generation that emerges in May and a nonoverwintering generation that emerges in July. However, it is possible that there is one generation per year with a long emergence period from May to July. Thus, these topics will be of interest for future research.

On the other hand, the number of individuals that were harvested in the B5 group in September was significantly fewer than the number of individuals that were harvested in the B5 group in July. As July is the irrigation season and the flow velocity is substantial even at the farm drain, there may be a few environments where a number of individuals that emerge in May can spawn in a consolidated paddy field in this study. Alternatively, because the individuals that have emerged in other waters come to the drainage canal system to spawn, the number of individuals in the B5 group will be high in September. In the future, the life history of this species in paddy fields will be clarified.

10.4.6   Effects of the Presence of Running Water in the Nonirrigation Season on Aquatic Animals in the Drainage Canal System

The effects running water in the nonirrigation season on aquatic animals are as follows: fishes can move to wintering places where water is deep, whereas the majority of aquatic insects, such as *Calopteryx atlata* and *Orthetrum albistyrum speciosum,* overwinter in the form of larvae in specific locations because their mobility is low. Thus, it is necessary to secure running water in the nonirrigation season to protect aquatic insects against drying. In general, the harvest of aquatic animals was substantial at the survey sites where there was flowing water in the nonirrigation season. In addition, it is notable that many *Orthetrum albistyrum speciosum* occurred at FD1, where the major water source was miscellaneous drainage and the water quality was poor.

Sympetrum infuscatum was identified in this study with 5 individuals confirmed at FD1, whereas 26 individuals were confirmed at FD2. As *Sympetrum infuscatum* has increased in recent years, the drying of paddy fields in the plains provides a relatively favorable environment state for this species, which originally tended to spawn in waterless places (Ueda 1998). In this study, *Sympetrum infuscatum* was considered to have selected the environment at FD2, where there was no flowing water in the nonirrigation season.

Daphnia pulex with some Rotifera and Protozoa have been documented to swarm over a short period after watering from May to July in the temporary waters of paddy fields (Mizuno and Urata 1964). Temporary waters, such as paddy fields, are used as breeding grounds for some fish in early irrigation. Then, plankton that occur in the temporary waters are used directly and indirectly by many larval and juvenile fishes or fishes that inhabit only the irrigation canals (Saitoh et al. 1988). These studies and the *Sympetrum infuscatum* identified in this area indicate that in addition to permanent waters, temporary waters that are dry in the nonirrigation season are important components of paddy ecosystems.

10.4.7 Proposals for Water Management in Consideration of Aquatic Animals

When considering water management that takes into account aquatic animals, running water is required in not only the main and lateral drains but also the farm drains in the nonirrigation season. However, the results of this study suggest that this approach of ensuring water flow in all farm drains detracts from paddy biodiversity. Thus, there is little need to ensure the flow rate in all farm drains in the nonirrigation season. Based on the average water depth at FD1, where *Orthetrum albistyrum speciosum* was caught, the depths in this study were approximately 5 cm in the nonirrigation season, with an almost stagnant water state. As the water temperature in the nonirrigation season was low in comparison with that in the irrigation season, fish in the drainage canal did not work during most of the nonirrigation season (Matsui and Satoh 2004b). Fish were less likely to move from main drains to farm drains in the nonirrigation season, and they were determined to be living somewhere in the main drain, where water was deep. Thus, it will be important to ensure that fish move to the main drain from the lateral and farm drains.

To ensure flowing water in the nonirrigation season, water must be taken from rivers and distributed from the irrigation canal system to the drainage canal system. At present, it is difficult to secure these new water rights because of the need to maintain the river flow in winter. Although water is considered to be rarely consumed, it is used and discharged into rivers (Satoh 2002). On the other hand, it is inevitable that the water flow rate is reduced in the river section from the intake point to the reduced water inflow point. Unless there is a particularly large impact that reduces water levels, both paddy and river ecosystems are rich because of the connection between the river and the agricultural drainage canal. Thus, having a small amount of water flowing in the nonirrigation season is preferable for the whole region. As a study on water passages, it is necessary to understand the ecological structure, including irrigation canals and paddy fields.

10.4.8   Proposals for Drainage Canal Maintenance Considering Aquatic Animals

(1)   **Drainage canal maintenance in consideration of fish habitat**

This drainage canal system serves as a tributary of rivers for *Opsariichthys platypus*. When fish such as *Opsariichthys platypus* can inhabit a river and proceed into a drainage canal system, the boundary of the rivers and paddy fields do not need to be clear. In wide areas of water between rivers and paddy fields, many fish that do not like temporary waters have been identified (Hirano et al. 2004). Therefore, the construction of a wetland between rivers and paddy fields will improve water quality and water levels. As a result, it would be easy for many fish to approach the drainage canal system from the river. In addition, wetlands can also be wintering sites in the nonirrigation season.

On the other hand, bottom fishes such as *Misgurnus anguillicaudatus* do not need to enter the paddy field drainage system from the river. It is desirable for them to ascend from wetlands to paddy fields to spawn. After spawning, they can descend and overwinter in the wetlands. However, the canal bed needs to be in a natural state so that the fish can overwinter in the mud in the farm drain. In fact, it is necessary to secure land to create wetlands. As the amount of abandoned farmlands has increased, it would be beneficial to create wetlands in these areas if possible (Hata 1987).

Notably, not only the wetland creation but also the continued existence the river and the drainage canal system; the main, lateral and farm drains in the drainage canal system; and year-round water flow are important for the conservation of biodiversity in paddy fields.

(2)   **Drainage canal maintenance in consideration of dragonfly habitat**

Calopteryx atlata and *Orthetrum albistyrum speciosum* use the lateral and farm drains, respectively, as the main spawning grounds and habitats in this drainage canal system. However, the larvae of both species that live in the drainage canal system from September to October showed a tendency to decrease after November. In particular, this trend was more remarkable in *Orthetrum albistyrum speciosum*.

Calopteryx atlata is a water-flowing species and shows a tendency to prefer the drainage canal, whereas *Orthetrum albistyrum speciosum* prefers still water and waters other than that in the drainage canal. Paddy fields will not be a habitat for these species because of drying in the nonirrigation season. Thus, the creation of the wetland described above is expected to be effective for conserving not only fishes but also dragonflies, especially *Orthetrum albistyrum speciosum*.

Box 10.1 Key Research Details

The field survey was conducted once a month for one year. As there are approximately 30 days in a month, which day is best for surveying? The answer is to follow the seasonal changes of the year: Spring Equinox (one day of March 20–21), Summer Solstice (one day of June 20–23), Autumnal Equinox (one day of September 22–24) and Winter Solstice (one day of December 21–23). These days are near the 20th of every month. Thus, it is appropriate to conduct a survey near the 20th every month to capture the seasonal changes in living things.

When you investigate a river, is it better to investigate from upstream or downstream? The answer is from upstream. The reason is as follows: Surveying rivers is more dangerous than surveying paddy fields. In particular, upstream areas are usually forest areas, so there are few people. Additionally, the sun sets early at upstream points. Therefore, it is important to start the survey upstream at an early time when you have the physical strength.

Determining the truth of things is simple. In complex cases, the truth is often not yet determined. Thus, in these cases, we must think further and pursue the truth. The process can be painful due to repeated trial and error, but there will always be breakthroughs. You must keep trying until you get there.

I would like young people to act not only in their own country but also in the world and contribute to the development and happiness of humankind. We are always born with some role (mission). The mission is different for each person. Therefore, you do not need to worry about others around you. Be aware your mission early and make it happen. That way, your life will surely be fulfilling and wonderful (Appendix Fig. 10.28).

Appendix

See Fig. 10.28.

Fig. 10.28 Children catching aquatic insects *(Photo* by Akira Matsui). *Notes* Photo date: April 5, 2020. The left is the author's eldest son, and the right is the author's second son

References

Arai Y (2001) The wonder of dragonflies (in Japanese). Doubutsusha Publishers Co., Ltd., Tokyo
Food, Agriculture and Rural Policy Council (2002) Guide of survey planning and designing to implement the project taking harmony with the environment into consideration (in Japanese). Ministry of Agriculture, Forestry and Fisheries, Tokyo
Fujioka M (1998) Crisis in rice fields herons warn (in Japanese). In: Ezaki Y, Tanaka T (eds) Conservation of waterfront environment—view point of biotic community. Asakura Publishing Co., Ltd., Tokyo, pp 34–52
Hasegawa M (1998) Frogs crowd depends on rice fields cultivation (in Japanese). In: Ezaki Y, Tanaka T (eds) Conservation of waterfront environment—view point of biotic community. Asakura Publishing Co., Ltd., Tokyo, pp 53–66

Hata K (1987) Improvement of canal in consideration of habitats for fishes (in Japanese). J Jpn Soc Irrigation Drainage Reclamation Eng 55:1067–1072. https://doi.org/10.11408/jjsidre1965.55.11_1067

Hirano T, Muraoka K, Yamashita S, Amano K (2004) Influence of the structure in the paddy field area on run-up of the fish (in Japanese). In: Ecology and civil engineering society 8th research workshop proceedings, pp 25–28

Hosoya K, Ashiwa H, Watanabe M, Mizuguchi K, Okazaki T (2003) *Zacco sieboldii*, a species distinct from *Zacco temminckii* (Cyprinidae). Ichthyol Res 50:1–8. https://doi.org/10.1007/s10 2280300000

Ishida S, Ishida K (1985) Odonata (in Japanese). In: Kawai T (ed) An illustrated book of aquatic insects of Japan. Tokai University Press, Tokyo, pp 33–124

Ishida S, Ishida K, Kojima K, Sugimura M (1988) Illustrated guide for identification of the Japanese Odonata (in Japanese). Tokai University Press, Tokyo

Kawai T (1985) An illustrated book of aquatic insects of Japan (in Japanese). Tokai University Publishing Division, Tokyo

Kawai T, Tanida K (2005) Aquatic insects of Japan: manual with keys and illustrations (in Japanese). Tokai University Publishing Division, Kanagawa

Kubota Z (1961) Ecology of the Japanese Loach, *Misgurnus anguillicaudatus* (CANTOR)—I. Ecological distribution (in Japanese with English Abstract): J Shimonoseki College Fisheries 11:141–176

Matsui A, Satoh M (2004a) Distribution of aquatic animals in the drainage systems created by paddy farmland consolidation in Shimodate City, Ibaraki Prefecture, Japan (in Japanese with English Abstract). Jpn J Conserv Ecol 9:153–163. https://doi.org/10.18960/hozen.9.2_153

Matsui A, Satoh M (2004b) A proposal for fish habitat improvement based on the analysis of fish distribution in the irrigation and drainage systems of a consolidated paddy field (in Japanese with English Abstract). Ecol Civ Eng 7:25–36. https://doi.org/10.3825/ece.7.25

Matsui A (2009) Growth of several fish and dragonfly species in the drainage system of a consolidated paddy field (in Japanese with English Abstract). Jpn J Conserv Ecol 14:3–11. https://doi.org/10.18960/hozen.14.1_3

Miyaji D, Kawanabe H, Mizuno N (1963) Colored illustrations of the freshwater fishes of Japan (in Japanese). Hoikusha Co., Ltd., Osaka

Mizuno N, Nakagawa N (1968) Growth of *Zacco platypus* (in Japanese). In: Mizuno N (ed) Ecology of the river and the fish in Osaka prefecture. Osaka Prefectural Fisheries Forestry Division, Osaka, pp 164–171

Mizuno T, Urata M (1964) The effect of the desiccation on plankton communities and the adaptability of plankton species for the drying-up of temporary pools (in Japanese). Physiol Ecol 12:225–229

Mori S, Nagoshi M (1989) *Zacco platypus* (in Japanese). In: Kawanabe H, Mizuno N, Hosoya K (eds) Freshwater fishes of Japan. Yama-kei Publishers Co., Ltd., Tokyo, pp 244–249

Moriyama H (1997) What happened is to preserve paddy fields?—From the viewpoint of biota (in Japanese). Rural Culture Association, Tokyo

Nakabo T (2000) Fishes of Japan with pictorial keys to the species (in Japanese). Tokai University Publishing Division, Tokyo

Nakabo T (2013) Fishes of Japan with pictorial keys to the species third edition (in Japanese). Tokai University Publishing Division, Kanagawa

Nakagawa S (2000) Consideration for ecology in the farm land consolidation (in Japanese). Rural Environ 16:48–53

Nakamura K (1952) Environment, food habit, spawning, development, growth and fisheries of *Zacco platypus* in Chikuma River (in Japanese with English Abstract). Bull Freshwater Fisheries Res Lab 1:2–25

Okuda S, Shibata T, Shimatani Y, Mizuno N, Yajima M, Yamagishi T (1996) *Opsariichthys platypus* (in Japanese). In: Japan RiverFront research Center (ed), Biological picture book of the river. Sankaido Publishing Co., Ltd., Tokyo, pp 332–333

Ozawa S (2000) The day in which *Oryzias latipes* disappear (in Japanese). Iwanamishoten Publishers Co., Ltd., Tokyo

Saitoh K, Katano O, Koizumi A (1988) Movement and spawning of several freshwater fishes in temporary waters around paddy fields (in Japanese with English Abstract). Jpn J Ecol 38:35–47. https://doi.org/10.18960/seitai.38.1_35

Satoh M (2002) Vision; To manage the region's water (in Japanese). J Jpn Soc Irrigation Drainage Reclamation Eng 70:797–798. https://doi.org/10.11408/jjsidre1965.70.9_797

Shinzawa K, Koide S (1963) Land readjustment of arable land (in Japanese). Iwanamishoten Publishers Co., Ltd., Tokyo

The Japanese Society of Irrigation, Drainage and Reclamation Engineering (2000) Agricultural engineering handbook basic edition (in Japanese). Japanese Society of Irrigation, Drainage and Reclamation Engineering, Tokyo

Tsuda M, Rokuyama M (1973) Aquatic insects, color nature guide 7 (in Japanese). Hoikusha Publishers Co., Ltd., Tokyo

Tsujii Y, Ueda T (2003) Ecological distribution of Medaka Fish, *Oryzias latipes* (Temminck et Schlegel), in concrete-lined ditches of paddy fields (in Japanese with English Abstract). Jpn J Environ Entomol Zool 14:179–192. https://doi.org/10.11257/jjeez.14.179

Ueda T (1998) The community of dragonfly in the paddy field (in Japanese). In: Ezaki Y, Tanaka T (eds) Conservation of waterside environment—from the viewpoint of biological community. Asakura Publishing Co., Ltd., Tokyo, pp 93–110

Chapter 11
Irrigation Canal System Survey

Abstract Based on the concept of establishing a continuous water flow and material cycle in a watershed, it is important to connect an irrigation canal system from the river to the drainage canal system and back to the former river. Even if biomass increases only in the drainage canal, the effect is limited. To enrich biodiversity in paddy fields, year-round water flows and natural materials in canal beds are important. However, information on the water volume that contributes to the environment and conserves paddy ecosystems is extremely limited. Therefore, we must address the technical issues related to water flow. The purpose of this study is to show the ecological significance of connecting irrigation and drainage canals in a consolidated paddy field and provide a practical plan based on a field survey of fish distributions at four sites in a paddy field in Shimodate City (now Chikusei City), Ibaraki Prefecture, Japan. The sampling interval was one week during May 2002 to June 2003. The results were the following: (1) the ten species of fishes caught in the canals were classified into two groups, fishes living mainly in drainage canals, such as *Misgurnus anguillicaudatus* and *Silurus asotus*, and those living in both irrigation and drainage canals, such as *Gnathopogon elongatus elongatus*. (2) The first group of fishes are distributed in the drainage canal system and move to paddy fields to facilitate movements between these two areas and expand their spawning grounds and habitats. These fishes are called 'drainage–paddy field fish.' The second group of fishes are distributed in both the irrigation and drainage canal systems, and it is important that movement is facilitated between the irrigation and drainage canal systems, which would also habitat. These fishes are called 'irrigation–drainage fish.' (3) An effective way to attain both fish conservation and high productivity is to employ a shallow farm drain by using an underdrain system for subsurface drainage. The small difference in elevation between farm drain and paddy plot allows the first group of fishes to enter the shallow water in the paddy field, and the connection of the main canal and shallow farm drain through water flow allows the second group of fishes to move between the irrigation and drainage canals. (4) As a fish conservation measure that can be conducted immediately, the fish that exist in the irrigation canal in the irrigation season can be moved to the shallow farm drain from the main canal before stopping the water flow for irrigation.

This chapter is a revised version of Matsui and Satoh (2004). Copyright 2004 Ecology and Civil Engineering Society, https://doi.org/10.3825/ece.7.25, Accessed November 28, 2021.

© The Author(s), under exclusive license to Springer Nature Singapore Pte Ltd. 2022
A. Matsui, *Wetland Development in Paddy Fields and Disaster Management*,
https://doi.org/10.1007/978-981-19-3735-4_11

Keywords Drainage canal · Farmland consolidation · Fish · Irrigation canal ·
Paddy field · Shallow farm drain

11.1 Introduction

From the viewpoint of conserving aquatic animals in Japanese rural areas, conventional agricultural infrastructure improvements and rural development projects have shown that water recycling is blocked by diversion weirs and dams in the rivers (Okuma 1994), aquatic animals cannot move independently in irrigation and drainage canals (Hata 1997) and aquatic animal habitats are being destroyed by drying in the canals in the nonirrigation seasons (Hata 1997). However, empirical studies on how decreasing water recycling impacts aquatic animals have rarely been carried out.

Based on the concept of establishing a continuous water flow and material cycle in a watershed, it is important to connect an irrigation canal system from the river to the drainage canal system and back to the former river. Even if biomass increases only in the drainage canal, the effect is limited. Because an irrigation canal is as long as the drainage canal in a paddy field, the drainage and irrigation canals serve the function of conserving aquatic animals by enhancing the space in paddy fields. To enrich biodiversity in paddy fields, year-round water flows and natural materials in canal beds are important (Matsui and Satoh 2004). However, information on the water volume that contributes to the environment and conserves paddy ecosystems is extremely limited. Therefore, we must address the technical issues related to water flow.

The purpose of this study is to show the ecological significance of connecting irrigation and drainage canals in a consolidated paddy field and provide a practical plan based on a field survey of fish distributions at four sites in a paddy field in Shimodate City (now Chikusei City), Ibaraki Prefecture, Japan (36° 21′ N, 139° 59′ E). The sampling interval was one week during May 2002 to June 2003. In addition, it should be noted that in this paper, the irrigation and drainage canal systems are each divided into the main, lateral and farm canals.

11.2 Methods

11.2.1 Survey Areas

The study site was the irrigation and drainage canals under the jurisdiction of Kawama Land Improvement District, Shimodate City (now Chikusei City), Ibaraki Prefecture, Japan. The area of Kawama Land Improvement District is located in the northeastern part of Shimodate City. The northern end of this area borders Tochigi Prefecture, and the eastern and western areas are between Kokai River and Gogyo River. This area is located in the paddy field zone and forms a slender area where the distance to the east

Table 11.1 Basic features of the survey sites

Survey sites		Year-round water flow	Bottom material	Side material	Canal width (m)	Canal depth (m)	Consolidation (year)
MC	Main canal	×[a]	Concrete	Concrete	1.50	1.38	1967
FC	Farm canal	×	Concrete	Concrete	0.25	0.25	1967
MD	Main drain	○	Concrete[b]	Concrete	3.20	1.20	1967
FD	Farm drain	×	Concrete[b]	Concrete	0.60	0.90	1967

[a]During the nonirrigation season, water flow was seen for approximately only two months in mid-winter from December 20, 2002 to February 28, 2003 to supply water for fire prevention
[b]Deposition of sand and mud was observed
Notes The survey was performed at two different canal levels each in irrigation and drainage canal systems
Source Modified from Matsui and Satoh (2004). Copyright 2004 Ecology and Civil Engineering Society

and west is approximately 2.5 km and that to the north and south is approximately 9 km. Villages are scattered throughout this area.

The irrigation water is taken from the Akaido diversion weir, which is located in the Gogyo River. The irrigation water runs through the main canal, lateral canal and farm canal in the irrigation canal system. Then, the irrigation water flows into the paddy field and runs through the farm drain, lateral drain and main drain in the drainage canal system. Ultimately, the drainage water is naturally drained into the Gogyo River.

The survey sites were located among the abovementioned irrigation and drainage canal systems, and generally, the four survey sites were set at the main canal, farm canal, farm drain and main drain (Table 11.1; Figs. 11.1, 11.2, 11.3, 11.4 and 11.5).

In addition, there is no clear distinction between the main and lateral drains and main and lateral canals. In this study, they were distinguished based on the difference in their physical environments, e.g., water depth and flow velocity. That is, in the drainage canal system, the sites located downstream of this area were considered the main drains, and the sites located upstream of this area were considered the lateral drains. Similarly, in the irrigation canal system, the sites located downstream of this area were considered the lateral canal, and the sites upstream of this area were considered the main canal. The farm canals and drains flowed directly beside the paddy fields.

The canal shapes were as follows: main canal width was 1.50 m, and main canal depth was 1.38 m; farm canal width was 0.25 m, and farm canal depth was 0.25 m; main drain canal width was 3.20 m, and main drain canal depth was 1.20 m; and farm drain canal width was 0.60 m, and farm drain canal depth was 0.90 m. The canal

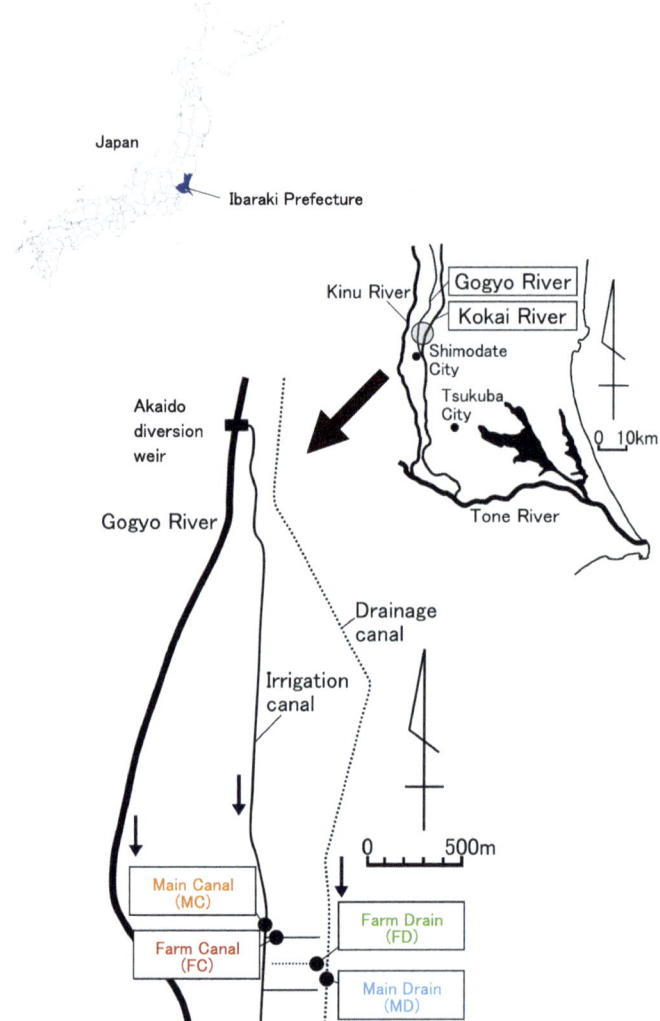

Fig. 11.1 Location of the survey sites. *Note* For information on the survey sites MC, FC, MD and FD, see Table 11.1. *Source* Modified from Matsui and Satoh (2004). Copyright 2004 Ecology and Civil Engineering Society

structure, which consisted of a canal bed and side materials, was concrete at all four sites.

There was year-round water flow in the main drain, while there was no flowing water in the nonirrigation season in the main canal, farm canal and farm drain. The irrigation season in this area was from the middle of April to the beginning of September. In addition, during the nonirrigation season, water flow occurred for only approximately two months in mid-winter from late December to February to supply

Fig. 11.2 Panoramic view of the main canal (*Photo* by Akira Matsui). *Notes* Photo date: May 2002. View from downstream to upstream

water for fire prevention in the main canal. Paddy farmland consolidation in this area began as an agricultural structure improvement project in 1964 and was completed in 1975. These four sites were established in 1967.

11.2.2 Survey Methods

In principle, sampling occurred at a weekly interval from May 20, 2002 to June 30, 2003.

1. **Physical environmental survey**

   Water depth, flow velocity, water temperature, hydrogen ion concentration (hereafter called the pH), dissolved oxygen (DO) concentration and electrical conductivity (EC) were measured. Water depth was measured by a meter stick. Flow velocity was surveyed by an electromagnetic portable current meter (manufactured by Nippon Hicon Co., LTD., Model 2000). Water temperature, pH, DO and EC were determined by a water quality meter (manufactured by DKK-TOA Corporation, Model WQC-22A). The water depth and the flow velocity were measured at the center of the farm canal and farm drain. These values were a mean of 3 measures, which were divided by 4 into the crossing direction at the main canal and main drain. In addition, the flow velocity was

Fig. 11.3 Panoramic view of the farm canal (*Photo* by Akira Matsui). *Notes* Photo date: May 2002. View from upstream to downstream

measured at a 60% water depth point. Two self-recording, pressure-type water-level gauges (manufactured by UIZIN Co., Ltd., data logger Model UIZ3635, pressure sensor Model UIZ-WL100) were installed approximately 1 km downstream at the main canal and main drain, and the water levels were recorded continuously every hour.

2. **Fish survey**

Fixed fishing nets installed at the main canal and main drain were large (diameter of bag net 0.4 m, length of bag net 2 m, length of sleeve net 6 m and opening size 5 mm), and those at the farm canal and farm drain were small (diameter of bag net 0.2 m, length of bag net 0.9 m, length of sleeve net 1 m and opening size 4 mm). The installation points of the fixed fishing nets at the survey sites are shown in Figs. 11.6 and 11.7. Installation of the fixed fishing nets was carried out toward an opening on the downstream side, and efforts were made to catch the fish running against the water flow. Because the fixed fishing nets at the main canal and main drain were not partitioned across the canals, the fish harvests included both the individuals that were descending from upstream and those ascending from downstream. On the other hand, as the fixed fishing nets at the farm canal and farm drain were partitioned across the canals, the fish harvests were limited to individuals that were ascending from downstream.

Fig. 11.4 Panoramic view of the main drain (*Photo* by Akira Matsui). *Notes* Photo date: May 2002. View from downstream to upstream

The duration of the fixed fishing nets surveys are indicated in Table 11.2. There was always running water from May 20, 2002 to June 30, 2003 at the main drain. The fixed fishing net was continuously used, and the fish were collected every week. The fixed fishing net was installed in succession during the irrigation season at the main canal, and the fish were collected every week. The irrigation water was suspended from September 5, 2002 to April 15, 2003 during the nonirrigation season. On the other hand, flowing water was temporarily restored to the main canal from December 20, 2002 to February 28, 2003, and the fixed fishing net was installed during that period. Because sizes of the fixed nets at the farm canal and farm drain were small compared to those at the main canal and main drain, dust was likely immediately caught in the nets, and running water was greatly inhibited by the fixed fishing nets. Therefore, fixed fishing nets were established for a period of four months from May to August 2002 and were placed for three days each month. The fish were collected every day. In addition, to investigate the ecology of the fish that used the paddy fields mainly as spawning grounds, the fixed fishing net was consecutively placed for three weeks from 9 to 30 June 2003 at the farm drain, and the fish were collected as frequently as twice a week. The harvested fish were counted on a monthly basis to understand seasonal changes in the number of fish caught at each survey site. The total numbers of fish in the harvests were shown as the number harvested in one week at the main canal and main drain and the number harvested in three days at the farm canal and farm drain.

Fig. 11.5 Panoramic view of the farm drain (*Photo* by Akira Matsui). *Notes* Photo date: May 2002. View from downstream to upstream

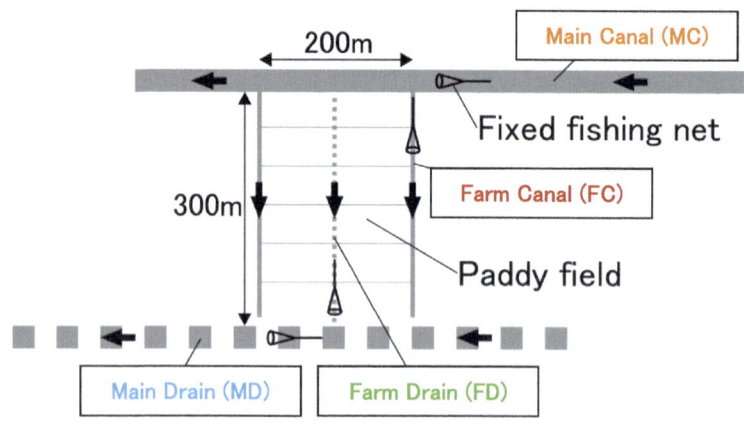

Fig. 11.6 Installation points of the fixed fishing nets in the irrigation and drainage canal systems. *Source* Modified from Matsui and Satoh (2004). Copyright 2004 Ecology and Civil Engineering Society

Fig. 11.7 Fixed fishing nets at the **a** main drain and **b** farm canal (*Photos* by Akira Matsui)

Table 11.2 Duration that fixed fishing nets were set at the survey sites

MC	MD	FC	FD
May 20, 2002–September 5, 2002	May 20, 2002–June 30, 2003	May 20, 2002–May 23, 2002	
December 20, 2002–February 28, 2003		June 17, 2002–June 20, 2002	
April 15, 2003–June 30, 2003		July 22, 2002–July 25, 2002	
		August 19, 2002–August 22, 2002	
		June 9, 2003–June 30, 2003[a]	
(254 days)	(406 days)	(FC: 12 days, FD: 33 days)	

[a]Only the farm drain was surveyed
Source Modified from Matsui and Satoh (2004). Copyright 2004 Ecology and Civil Engineering Society

The collected fish, except for large fish, were immediately fixed with a 10% formalin aqueous solution in the field. After returning to the laboratory, the fish species were identified according to Nakabo (2000), and the standard length and wet weight were measured. The measurement of the wet weight was performed by using an electronic balance (manufactured by Sartorius Japan K. K., Model BP1200). The large fish were released after the standard length and wet weight were measured in the field. It should be noted that the minimum units of the standard length and wet weight were 0.1 cm and 0.01 g, respectively.

Despite the absence of water flow in the nonirrigation season at the main canal, there were three pools. These pools were maintained at the intersection of the irrigation canal and farm road and were formed by digging a canal bed with a depth of approximately 1.5 m. Each pool was approximately 2 m^2. The pools were surveyed with a spoon net (bottom 0.35 m, net height 0.30 m and opening size 3 mm) in October, November and December 2002 and March 2003. The spoon net was placed five times at each pool. The collected fish, except for large fish, were immediately fixed with a 10% formalin aqueous solution in the field. After returning to the laboratory, the fish species were identified according to Nakabo (2000), and the standard length and wet weight were measured. The large fishes were released after the standard length and wet weight were measured in the field.

11.3 Results

11.3.1 Physical Environment in the Irrigation and Drainage Canal Systems

In terms of the hydrological conditions of the survey area, seasonal changes in precipitation and water depth at the main drain and main canal are shown in Fig. 11.8. The precipitation was recorded by the Automated Meteorological Data Acquisition System (AMeDAS) in Kyowa Town near Shimodate City (now Chikusei City). The total precipitation from April 2002 to March 2003 was 990 mm, which was close to the average value of 1167 mm over the past 20 years. The water at the main drain was generally deep in the irrigation season and fluctuated greatly according to rainfall, rising to 1.00 m from June to July 2002 due to flooding. On the other hand, the water depth at the main canal fluctuated in the irrigation season compared to that at the main drain. A water volume of approximately half that in the irrigation season temporarily flowed at the main canal in the nonirrigation season.

The seasonal changes in the water depth and flow velocity based on sampling every week at the survey sites are shown in Fig. 11.9. The trends in the water depth and flow velocity at each survey site were substantial in the irrigation season and minimal in the nonirrigation season. There was year-round water flow at the main drain and seasonal water flow at the main canal, farm canal and farm drain. However, the water flow at the main canal was seen for only approximately two months in midwinter from December 2002 to February 2003. In general, seasonal changes in the water depth and flow velocity in the irrigation canal system were larger than those in the drainage canal system. In addition, the water depths at the farm canal and farm drain were low because the water continuously flowed from middle to late June 2002.

The seasonal changes in water temperature based on sampling every week at the survey sites are presented in Fig. 11.10. The maximum water temperatures were 26.0 °C at the main canal, 25.8 °C at the farm canal, 25.6 °C at the main drain and 28.0 °C at the farm drain in July 2002. The minimum water temperatures were 14.3 °C at the farm canal and 14.1 °C at the farm drain in April 2003 because there was no water flow in winter, while the minimum temperatures were 5.5 °C at the main canal and 5.4 °C at the main drain in January 2003.

The seasonal changes in pH and EC based on sampling every week at the survey sites are shown in Fig. 11.11. The pH at the main drain was remarkably high from October to April in the nonirrigation season. It tended to exceed 6.0–7.5, which is a water quality standard of agricultural water (The Japanese Society of Irrigation, Drainage and Reclamation Engineering 2000). The EC at the main drain was significantly high from November to February in the nonirrigation season. It tended to exceed 30 mS/m, which is a water quality standard of agricultural water (The Japanese Society of Irrigation, Drainage and Reclamation Engineering 2000).

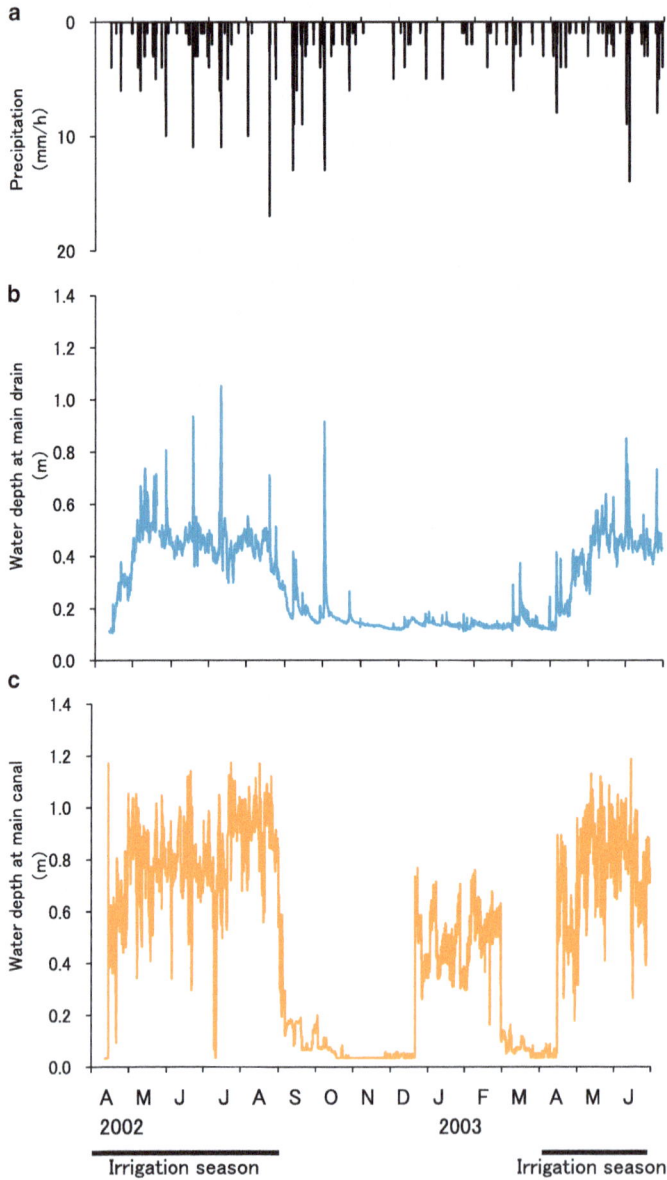

Fig. 11.8 Seasonal changes in **a** precipitation and water depth at the **b** main drain and **c** main canal. *Notes* The water at the main drain was generally deep in the irrigation season from May to August, and it fluctuated greatly according to rainfall. In the nonirrigation season, water flow at the main canal occurred only for approximately two months in mid-winter to supply water for fire prevention. *Source* Modified from Matsui and Satoh (2004). Copyright 2004 Ecology and Civil Engineering Society

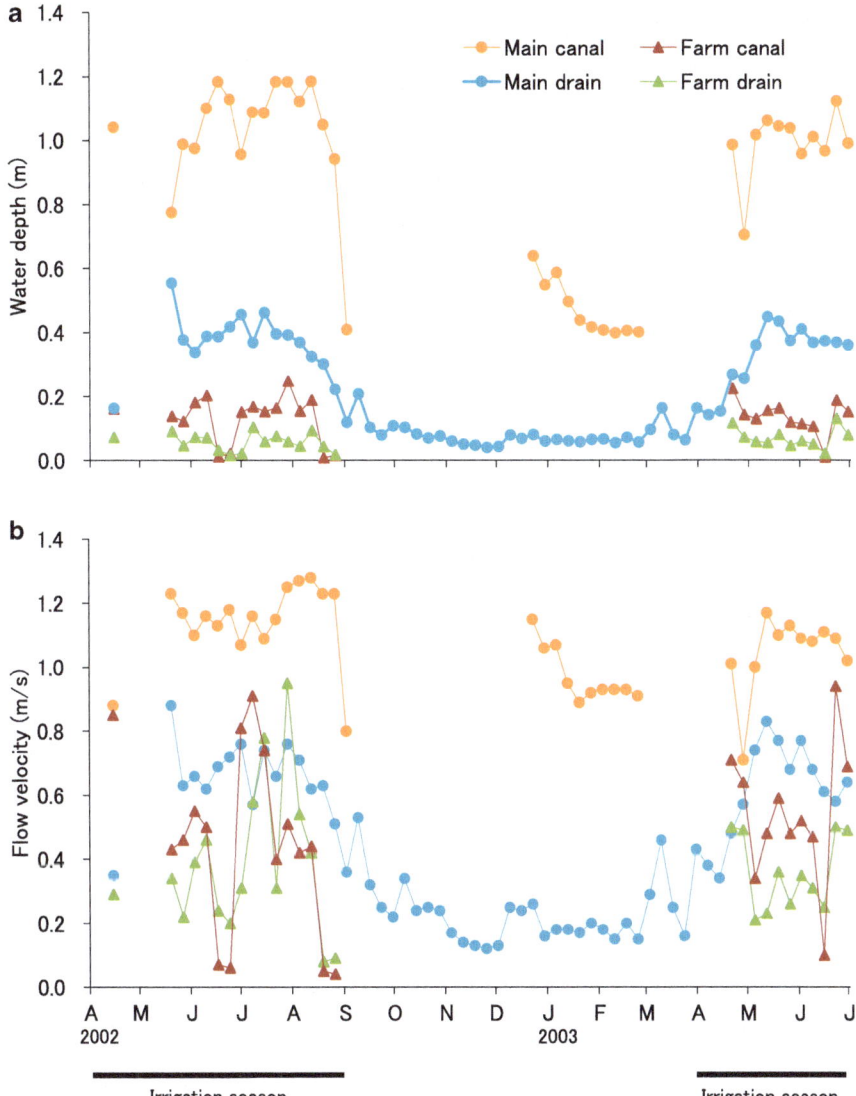

Fig. 11.9 Seasonal changes in **a** water depth and **b** flow velocity based on sampling every week at the survey sites. *Note* There was year-round water flow at the main drain, while seasonal water flow occurred at the main canal, farm canal and farm drain. *Source* Modified from Matsui and Satoh (2004). Copyright 2004 Ecology and Civil Engineering Society

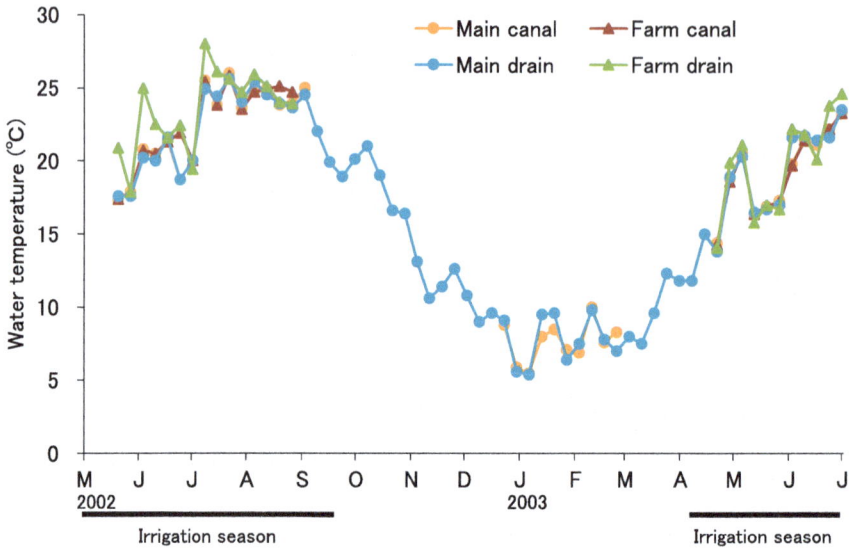

Fig. 11.10 Seasonal change in water temperature based on sampling every week at the survey sites. *Note* There was year-round water flow at the main drain, while seasonal water flow occurred at the main canal, farm canal and farm drain. *Source* Illustration based on Matsui and Satoh (2004)

11.3.2 Biomass of Fish in the Irrigation and Drainage Canal Systems

The number of fish caught at the survey sites in the irrigation and drainage canal systems from May 2002 to June 2003 by setting fixed fishing nets was 1141 individuals in total. The species were as follows: *Cyprinus carpio*, *Carassius* sp., *Opsariichthys platypus*, *Candidia temminckii*, *Tribolodon hakonensis*, *Gnathopogon elongatus elongatus*, *Pseudogobio esocinus esosinus*, *Misgurnus anguillicaudatus*, *Silurus asotus* and *Rhinogobius* sp. (Table 11.3). The order of the listed fishes follows Nakabo (2000). Identification of *Candidia temminckii* is based on Hosoya et al. (2003).

Box plots and statistical results for fish biomass at the survey sites are shown in Fig. 11.12 and Table 11.4. As a result of the Kruskal–Wallis test, *Misgurnus anguillicaudatus* and *Silurus asotus* showed a significant difference in their biomass values. The Steel–Dwass test for *Misgurnus anguillicaudatus* showed a significant difference in biomass values between MD and MC, MD and FC, MC and FD and FD and FC. In addition, the Steel–Dwass test for *Silurus asotus* showed a significant difference in biomass values between MD and MC.

Based on the seasonal changes in the number of fish caught at the survey sites, more fish were caught in the irrigation season, and less fish were caught in the nonirrigation season (Fig. 11.13). Because the fixed fishing net installation duration was different at survey sites, the numbers could not be compared simply. Overall,

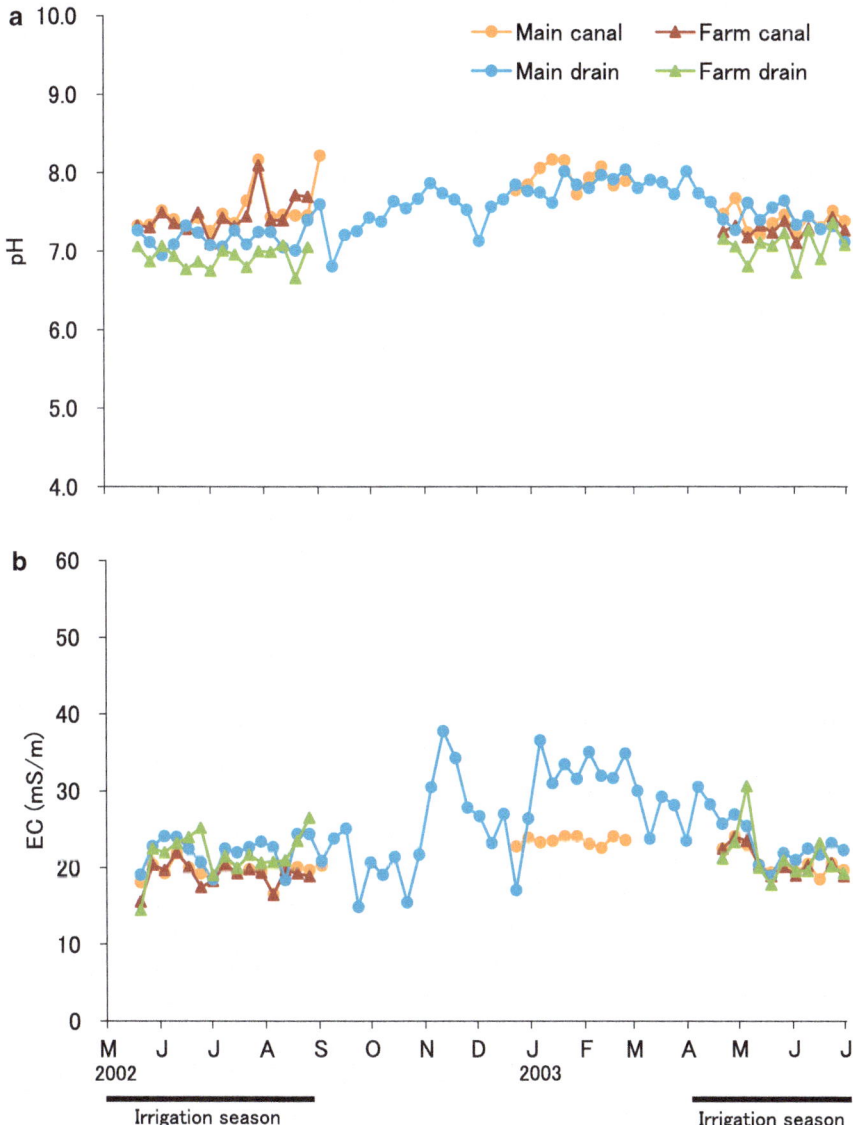

Fig. 11.11 Seasonal changes in **a** pH and **b** EC based on sampling every week at the survey sites. *Note* There was year-round water flow at the main drain, while seasonal water flow occurred at the main canal, farm canal and farm drain. *Source* Illustration based on Matsui and Satoh (2004)

Table 11.3 Total numbers of fish caught at the survey sites

Species	Irrigation system		Drainage system		Total
	MC	FC	MD	FD	
Cyprinus carpio	5	3	11	13	32
Carassius sp.	5	0	4	2	11
Opsariichthys platypus	0	0	4	0	4
Candidia temminckii	3	0	0	0	3
Tribolodon hakonensis	14	1	11	0	26
Gnathopogon elongatus elongatus	180	17	247	18	462
Pseudogobio esocinus esocinus	4	0	2	0	6
Cyprinidae	24	0	12	0	36
Misgurnus anguillicaudatus	7	0	109	315	431
Silurus asotus	7	0	117	5	129
Rhinogobius sp.	0	0	1	0	1
Total	249	21	518	353	1141

Notes The fixed fishing nets were set for 254 days at the main canal (MC), 12 days at the farm canal (FC), 406 days at the main drain (MD) and 33 days at the farm drain (FD). The values show the actual number of fish caught at the survey sites. The order of the listed fishes follows Nakabo (2000)

Source Modified from Matsui and Satoh (2004). Copyright 2004 Ecology and Civil Engineering Society

however, the number of fish caught at the main canal, main drain and farm drain was high, while that at the farm canal was conspicuously low. Only one individual of *Tribolodon hakonensis* was caught at the farm canal during the period of temporary water flow in the nonirrigation season from 23 to 30 December 2002.

On the other hand, 109 individual fish were harvested at three pools in the main canal that were maintained in the nonirrigation season (Table 11.5; Fig. 11.14). The species were as follows: *Candidia temminckii, Hemibarbus barbus, Squalidus chankaensis biwae, Misgurnus anguillicaudatus* and *Rhinogobius* sp. The fish living at the pools were surveyed with a spoon net in October, November and December 2002 and March 2003. Many juveniles of Cyprinidae were caught in October, November and December 2002 before intake. On the other hand, *Candidia temminckii* and *Squalidus chankaensis biwae* were harvested in March 2003 after intake ended.

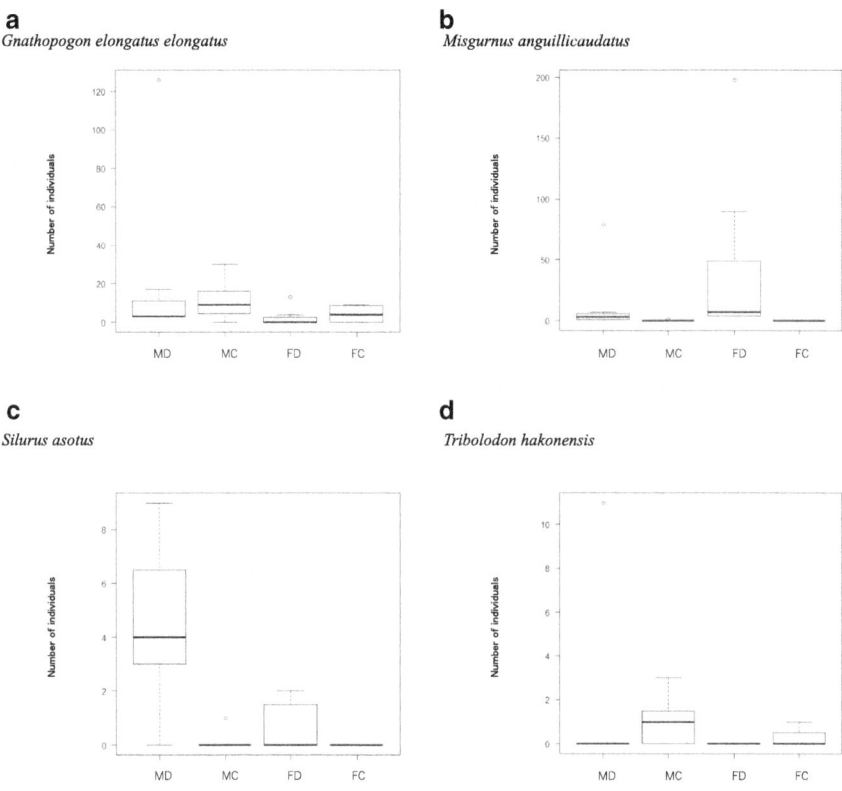

Fig. 11.12 Box plot of fish biomass at the survey sites. *Source* Illustration based on Matsui and Satoh (2004)

11.4 Discussion

11.4.1 Physical Environment in the Irrigation and Drainage Canal Systems

The seasonal changes in water depth and flow velocity fluctuated more in the irrigation canal system than in the drainage canal system. Hence, it is likely difficult for fish to inhabit the irrigation canal system. Thus, the anthropogenic impact on the irrigation canal system is larger than that on the drainage canal system because the irrigation water is supplied for the purpose of irrigation. In other words, the flow rate in the irrigation canal system was significantly increased by intake from the Akaido diversion weir to flood paddy fields. On the other hand, there was no flowing water in the irrigation canal system in the nonirrigation season because the irrigation water was stopped. When the rainfall was substantial in the irrigation season, irrigation water was adjusted so that the water did not overflow from the irrigation canal.

Table 11.4 Statistical results for fish biomass at the survey sites

a					b				
Gnathopogon elongatus elongatus					*Misgurnus anguillicaudatus*				
Kruskal–Wallis test *P* value = 0.145					Kruskal–Wallis test *P* value = 0.000				
Steel–Dwass test					Steel–Dwass test				
	MD	MC	FD	FC		MD	MC	FD	FC
MD		0.988	0.178	0.865	MD		*0.010*	0.234	*0.031*
MC			0.230	0.599	MC			*0.005*	0.874
FD				0.976	FD				*0.031*
FC					FC				

c					d				
Silurus asotus					*Tribolodon hakonensis*				
Kruskal–Wallis test *P* value = 0.003					Kruskal–Wallis test *P* value = 0.114				
Steel–Dwass test					Steel–Dwass test				
	MD	MC	FD	FC		MD	MC	FD	FC
MD		*0.026*	0.071	0.080	MD		0.583	0.749	0.992
MC			0.572	0.874	MC			0.111	0.660
FD				0.469	FD				0.548
FC					FC				

Note A italic values indicates a significant difference
Source Illustration based on Matsui and Satoh (2004)

In contrast, there was year-round water flow in the drainage canal system, and the seasonal changes in the water depth and flow velocity at the main drain were small in comparison with those at the main canal.

Regarding the seasonal changes in the water depth at the main drain, the maximum value was 1.05 m on July 11, 2002. Most of the year, the water depth was less than 0.50 m. Given that the difference in elevation between the paddy field and farm drain in this survey area was approximately 1.00 m, the frequency by which the drain water connected with the paddy water was low; thus, the irrigation and drainage canal systems were considered to have been generally separate.

11.4.2   Features of the Fish in the Irrigation and Drainage Canal Systems

The characteristics of the fish caught in the irrigation and drainage canal systems in this survey area were as follows: the biomass of *Misgurnus anguillicaudatus* and *Silurus asotus* was large in the drainage canal system and significantly small in the irrigation canal system. On the other hand, the biomass of *Gnathopogon elongatus elongatus* was large in both irrigation and drainage canal systems. The biomass of

Fig. 11.13 Seasonal changes in the number of fish caught at the survey sites. *Notes* **a** Main canal, **b** farm canal, **c** main drain and **d** farm drain. *Gnathopogon elongatus elongatus*, *Misgurnus anguillicaudatus* and *Silurus asotus* were the major species. Other species included *Cyprinus carpio*, *Carassius* sp., *Opsariichthys platypus*, *Candidia temminckii*, *Tribolodon hakonensis*, *Pseudogobio esocinus esocinus* and *Rhinogobius* sp. *Source* Modified from Matsui and Satoh (2004). Copyright 2004 Ecology and Civil Engineering Society

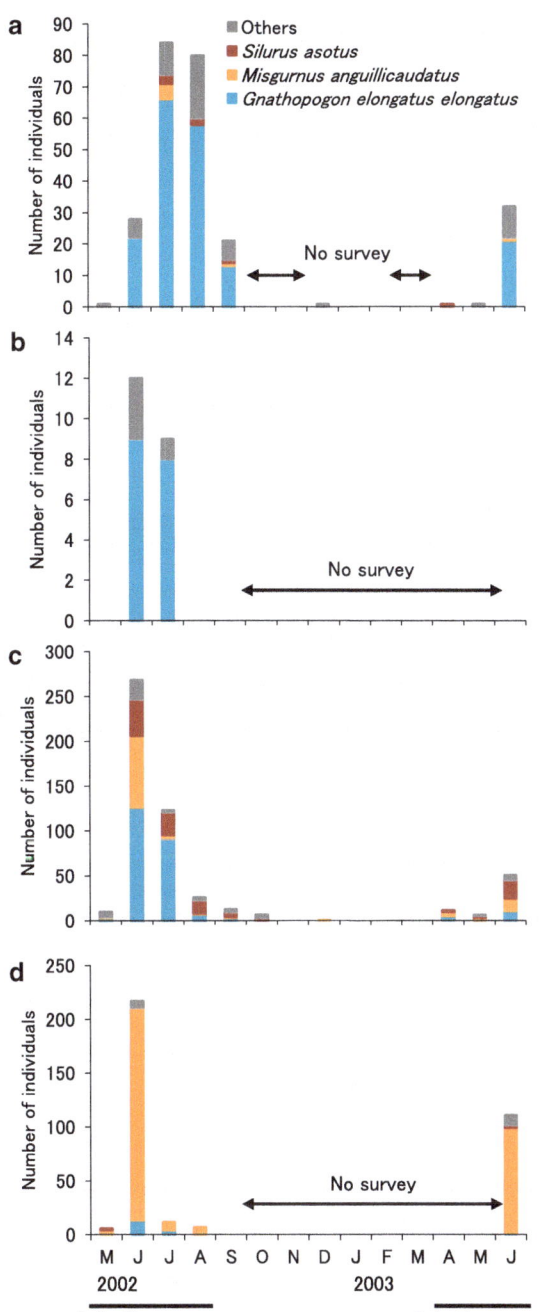

Table 11.5 Total numbers of fish caught at the pools in the main canal in the nonirrigation season

Species	2002			2003	Total
	Oct.[a]	Nov.[a]	Dec.[a]	Mar.[b]	
Candidia temminckii	0	0	0	2	2
Hemibarbus barbus	0	1	0	0	1
Squalidus chankaensis biwae	0	0	0	1	1
Cyprinidae	40	15	27	0	82
Misgurnus anguillicaudatus	0	4	2	0	6
Rhinogobius sp.	0	6	10	1	17
Total	40	26	39	4	109

[a] Before intake
[b] After intake ended
Notes There were three pools, and the size of each pool was approximately 2 m². A spoon net was placed into a pool five times. The values show the actual number of fish caught at the pools. The order of listed fishes follows Nakabo (2000)
Source Reprinted from Matsui and Satoh (2004). Copyright 2004 Ecology and Civil Engineering Society

Cyprinus carpio, *Carassius* sp., *Tribolodon hakonensis* and Cyprinidae was small; however, they were caught in both the irrigation and drainage canal systems.

Misgurnus anguillicaudatus and *Silurus asotus* were not caught in the irrigation canal because *Misgurnus anguillicaudatus* lives in paddy fields, wetlands and their surroundings (Saitoh 1989) and *Silurus asotus* lives in lakes and the middle and lower reaches of rivers (Kobayakawa 1989); thus, both fish prefer to be in the drainage canal with a still water environment. The irrigation canal with a flowing water environment and large seasonal changes in the water depth and flow velocity is not a habitat for these two species. The spawning season of *Misgurnus anguillicaudatus* and *Silurus asotus* is from May to July, and the spawning grounds are paddy fields (Council of Food, Agriculture and Rural Area Policies 2002). That is, *Misgurnus anguillicaudatus* and *Silurus asotus*, which are mainly distributed in the drainage canal, are estimated to move from the drainage canal to the paddy fields for spawning.

On the other hand, *Gnathopogon elongatus elongatus* spawns in low water flows, irrigation canals and paddy fields (Hosoya 1989) and buries its eggs in the vegetation and riverbeds to reduce larval mortality (Saitoh et al. 1988). Hence, this species can inhabit the irrigation canal system as well as the drainage canal system.

Of the total number of fish caught at the survey sites, *Gnathopogon elongatus elongatus* was the most frequently caught, followed by *Misgurnus anguillicaudatus* and *Silurus asotus* in decreasing order. The fish harvests were substantial from June to July 2002 and June 2003 at the main drain, where there was year-round water flow and the water temperature in June was approximately 20 °C. The main spawning season of *Gnathopogon elongatus elongatus* was presumed to be from May to June, when the water temperature is near 20 °C (Yata 1979). *Misgurnus anguillicaudatus* hides in the mud in winter. When the mud temperature at a depth of approximately 10 cm

Fig. 11.14 a Dry conditions and **b** pools in the main canal (*Photos* by Akira Matsui)

reaches more than approximately 13 °C, *Misgurnus anguillicaudatus* moves from the mud to the water after wintering (Kubota 1961). Thus, the relationship between the ecology of this fish and the water temperature is close. The water temperature at the main drain in June is approximately 20 °C and is favorable as fish habitat. Therefore, the number of individuals caught at the survey sites was generally high in the irrigation season and low in the nonirrigation season because the fish scarcely moved as the water temperature in the nonirrigation season was low in comparison with that in the irrigation season.

As noted above information, the fishes in the irrigation and drainage canal systems of a consolidated paddy field were classified into two groups: fishes living mainly in the drainage canal system (drainage–paddy field fishes), such as *Misgurnus anguillicaudatus* and *Silurus asotus*, and those living both in the irrigation and drainage canal systems (irrigation–drainage fishes), such as *Gnathopogon elongatus elongatus*.

11.4.3 Fishes at the Pools of the Irrigation Canal System

The characteristics of the fish caught at the pools in the irrigation canal system in the nonirrigation season were as follows: Cyprinidae juveniles were the most frequently caught fish likely because the individuals were spawning and hatching from spring to summer. On the other hand, in the nonirrigation season, given the flowing water, Cyprinidae juveniles were swept downstream. Of the Cyprinidae, *Candidia temminckii* and *Squalidus chankaensis biwae* were sourced from the Gogyo River.

The survey was part of the National Census on River Environments and carried out at six sites in the Kokai River in May and October 1995. The result was that the total number of fish caught during the survey was 1193 individuals (Ministry of Construction River Bureau River Environment Division 1997). Among the fish caught, there were 150 *Candidia temminckii* (12.6% of the total fish) and 69 *Squalidus chankaensis biwae* (5.8% of the total fish). Therefore, the *Candidia temminckii* and *Squalidus chankaensis biwae* caught at the pools can also live in the Gogyo River, which is a tributary of the Kokai River. These fish at the pools in the irrigation canal system were determined to be individuals that flowed downstream from the Gogyo River.

11.4.4 Issues with Conventional Irrigation and Drainage Canal Maintenance

When focusing on the issues with conventional irrigation and drainage canal maintenance, the fish within the canals can be divided into two types: drainage–paddy field fish and irrigation–drainage fish. It is desirable for the drainage–paddy field fish to be able to move from the drainage canal system to the paddy field to spawn. However, in the present situation, the difference in elevation between farm drains and paddy plots is approximately 1.00 m; thus, movement of the fish seems to be difficult. For the irrigation–drainage fish, it is preferable that they are able to move between the irrigation and drainage canal systems.

Three patterns can be considered when developing methods to connect the irrigation and drainage canal systems: (1) farm canal → paddy field → farm drain, (2) main canal → farm drain and (3) farm canal → main drain. However, since the difference in the elevation of the canal connection points is large, fish movement is difficult.

As described above information, regarding the effects of conventional irrigation and drainage canal maintenance on fish, the drainage–paddy field fish and irrigation–drainage fish, the difference in elevation between the irrigation and drainage canal systems is a major problem in consideration of fish ecology.

11.4.5 Conservation Measures for Drainage–Paddy Field Fish

Methods to maintain irrigation and drainage canal maintenance will need to pay specific attention to the fish that are distributed in the drainage canal system and move to paddy fields, such as *Misgurnus anguillicaudatus* and *Silurus asotus*, to facilitate movements between these two areas and expand their spawning grounds and habitats.

Establishing small fishways has been proposed to eliminate the difference in elevation between paddy fields and drainage canals (Hata 2000). However, the cost of installing a fishway on each individual paddy field is a large constraint.

As an environmentally friendly measure, Hirota (2003) reported a case of adopting a two-story at the farm drain. The farm consolidation scheme to maintain farmlands and save labor for slope management (Shinzawa and Koide 1963) should be reevaluated from the modern ecological conservation perspective. The proposal to employ a shallow farm drain by installing a culvert for underground water drainage would result a variety of physical and economic benefits.

Constructing a shallow farm drain in the earthen canal would ensure continuity between the drainage canal system and the paddy field and would expand the fish habitat. The area of the earthen canal is also small in this case; thus, the burden of maintaining the earthen canal would need to be determined. Moreover, the high productivity of the paddy field would be supported by an underground water drainage system with a pipeline (Fig. 11.15). By adopting this method, the water surface between the shallow farm drain and paddy field would be continuous at the time of heavy rain. This scenario would provide the environmental conditions under which *Misgurnus anguillicaudatus* and *Silurus asotus* can ascend easily from the shallow farm drain to the paddy field. It would not be necessary to install a fishway on each

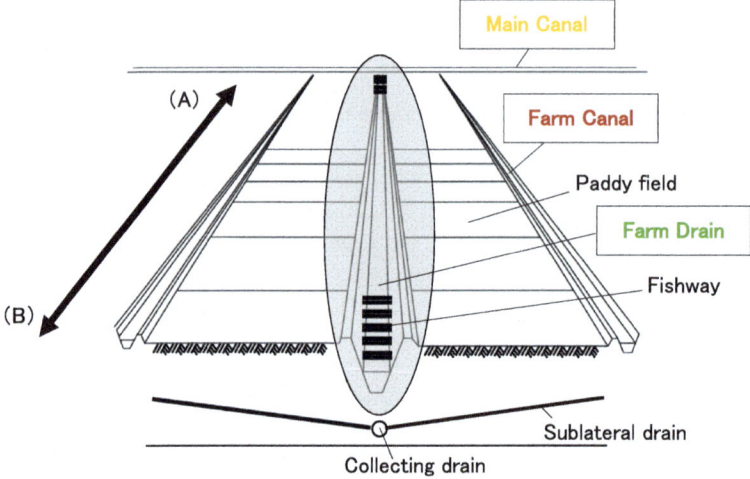

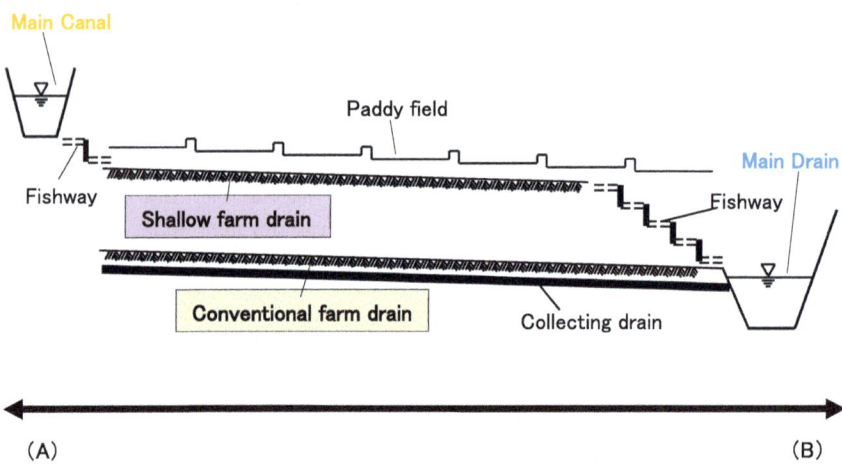

Fig. 11.15 Proposal of a field system connecting a drainage canal to a paddy plot and irrigation canal. *Notes* Conventional deep farm drain connects both a surface and subsurface drain. Introduction of a shallow farm drain provides the fish in the farm drain with extensive accessibility to the paddy plot and main canal. The subsurface water should be drained with the underdrain system separately from the surface water. *Source* Modified from Matsui and Satoh (2004). Copyright 2004 Ecology and Civil Engineering Society

paddy field; therefore, the number of fishways could be reduced significantly. On the other hand, the construction cost would increase due to the adoption an underground water drainage system compared with the drainage associated with conventional paddy farmland consolidation. However, the construction cost would be reduced by placing the shallow farm drains at predetermined intervals.

11.4.6   Conservation Measures for Irrigation–Drainage Fish

For fish, such as *Gnathopogon elongatus elongatus*, to be distributed in both the irrigation and drainage canal systems, it is important that movement is facilitated between the irrigation and drainage canal systems, which would also habitat. As shown in Fig. 11.15, it is proposed that fishways be installed at two places in the most upstream area and the most downstream end of the shallow farm drain. The most upstream area is in contact with the main canal. As there is no large difference in the water level between the main canal and the shallow farm drain, it is easy to connect them. On the other hand, the most downstream end is in contact with the main drain. Because there is a large difference in the water level between the shallow farm drain and the main drain, it is necessary to install many fishways at this end. By providing a certain flow rate in the irrigation season, it is believed that *Gnathopogon elongatus elongatus* would be able to move between the irrigation and drainage canal systems.

Furthermore, since 109 individual fish were caught at the pools in the main canal in the nonirrigation season, it would be counterproductive for the pools to be used as wintering places in the irrigation canal system. However, if it is possible for fish to descend in the drainage canal system in the nonirrigation season, then this scenario seems more preferable from the standpoint of expanding the habitat area and securing food resources. Based on these conclusions, it is extremely important to connect the main canal and the shallow farm drain.

11.4.7   Step-By-Step Measures for Fish Conservation

As the irrigation and drainage canals are separate, the fish that flow into the irrigation canal from the river in the irrigation season die in the canal in association with the cessation of irrigation water. In terms of fish conservation, the situation described above is a great problem. To address this problem, it is desirable to adopt conservation measures, as shown in Fig. 11.15. As a fish conservation measure that can be conducted immediately, the fish that exist in the irrigation canal in the irrigation season can be moved to the shallow farm drain from the main canal before stopping the water flow for irrigation.

In the district where there is no flowing water in the drainage canal in the nonirrigation season, there are no wintering places for fishes. Hence, it is necessary to provide fish conservation areas where the canal bottom is deep and the water flow is maintained in winter. It is proposed that a wetland downstream of the main drain be constructed (Hata 2000).

In addition, it is necessary to discharge water into the drainage canal from the irrigation canal to connect the main canal and the shallow farm drain with flowing water in the irrigation season. This is an invalid discharge for irrigation. However, as the discharge flows down the drainage channel without being consumed, there would be no water consumption problems related to reusing the water as irrigation water at the appropriate places (Satoh et al. 1998).

Based on this study, it was found that many fish lived in the consolidated paddy fields. However, if paddy fields are dried and implement crop rotations to ensure high productivity of agriculture, then the fish habitats quickly decrease in quality. Moving forward, it will be important to ensure that farmland is consolidated while considering the regional environment, and the improvement plans presented in this chapter will help in implementing farmland consolidation.

Box 11.1 Irrigation and Drainage Systems

In the past, both irrigation and drainage canals were used, but now irrigation and drainage canals are separate (Appendix Fig. 11.16). This approach is called paddy farmland consolidation. A consolidated paddy field improves rice productivity, and groundwater levels can be controlled by field maintenance. Paddy farmland consolidation has succeeded in converting wet fields to dry fields, and large agricultural machinery can be brought onto the paddy fields.

However, conventional paddy farmland consolidation in Japan, which aims to increase farming efficiency by improving the drainage conditions of paddy fields and the independent creation of irrigation and drainage canals, has been said to have negative impacts on biodiversity in rural areas.

Wetlands are very important for many organisms, and Japanese rice fields are representative wetlands. In recent years, the biological conservation function of rice fields has been reviewed, and the shallow farm drain and fish-retreat ditch introduced in this book have been constructed.

Appendix

See Fig. 11.16.

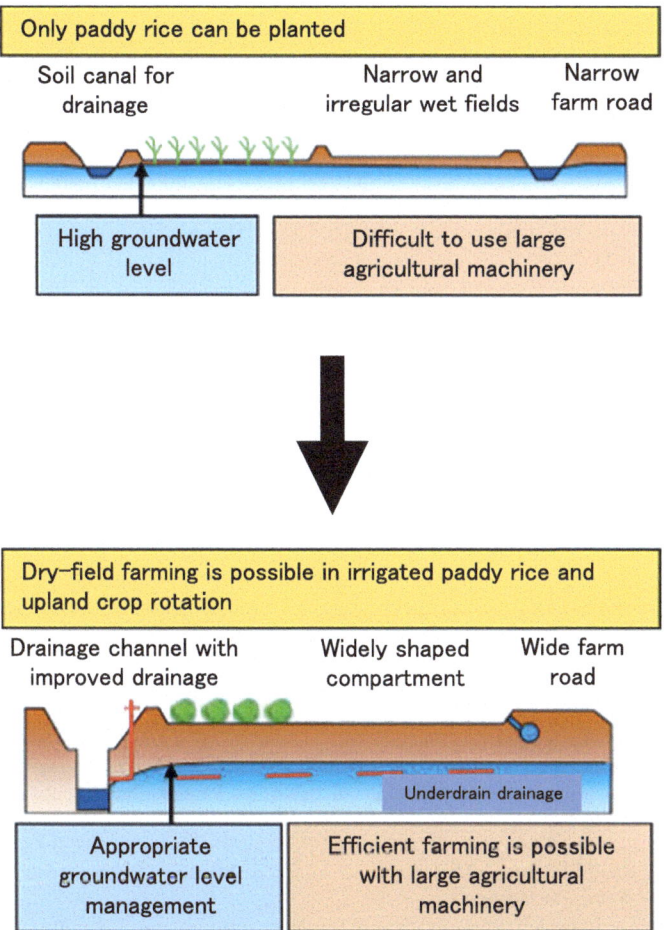

Fig. 11.16 Change in irrigation and drainage systems. *Source* Modified from the Yamaguchi Prefecture, https://www.pref.yamaguchi.lg.jp/cms/a17500/nouson/kakujigyo.html, Accessed November 3, 2021 (in Japanese)

References

Council of Food, Agriculture and Rural Area Policies (2002) Handbook of survey planning and designing in consideration of environment. Ministry of Agriculture, Forestry and Fisheries, Tokyo (in Japanese)

Hata K (1997) Current situation and problems of irrigation and drainage systems. In: Advice Center for Rural Environment Support (ed) Improvement methods of agricultural irrigation and drainage facilities in consideration of biota conservation. Advice Center for Rural Environment Support, Tokyo (in Japanese)

Hata K (2000) Run-up of the fish to the paddy field. Rural Environ 16:61–69 (in Japanese)

Hirota J (2003) Issues surrounding the farmland development and environment-friendly, multi-functionality. Res Agric Low 38:42–54 (in Japanese)

Hosoya K (1989) *Gnathopogon elongatus elongatus*. In: Kawanabe H, Mizuno N, Hosoya K (eds) Freshwater fishes of Japan. Yama-kei Publishers Co., Ltd., Tokyo, pp 298–299 (in Japanese)

Hosoya K, Ashiwa H, Watanabe M, Mizuguchi K, Okazaki T (2003) *Zacco sieboldii*, a species distinct from *Zacco temminckii* (Cyprinidae). Ichthyol Res 50:1–8. https://doi.org/10.1007/s10 2280300000

Kobayakawa M (1989) *Silurus asotus*. In: Kawanabe H, Mizuno N, Hosoya K (eds) Freshwater fishes of Japan. Yama-kei Publishers Co., Ltd., Tokyo, pp 412–415 (in Japanese)

Kubota Z (1961) Ecology of the Japanese loach, *Misgurnus anguillicaudatus* (CANTOR)—I. Ecological distribution. J Shimonoseki Coll Fish 11:141–176 (in Japanese with English abstract)

Matsui A, Satoh M (2004) A proposal for fish habitat improvement based on the analysis of fish distribution in the irrigation and drainage systems of a consolidated paddy field. Ecol Civ Eng 7:25–36 (in Japanese with English abstract). https://doi.org/10.3825/ece.7.25

Ministry of Construction River Bureau River Environment Division (1997) Census Yearbook of River waterfront river version fish and benthic animal research edition in 1995. Sankaido Publishing Co., Ltd., Tokyo (in Japanese)

Nakabo T (2000) Fishes of Japan with pictorial keys to the species. Tokai University Publishing Division, Tokyo (in Japanese)

Okuma T (1994) Connection with river and people. In: Yokohama Environmental Science Research Institute (ed) Association for Nature Restoration and Conservation 10th anniversary symposium material book. Association for Nature Restoration and Conservation, Shizuoka, pp 92–95 (in Japanese)

Saitoh K (1989) *Misgurnus anguillicaudatus*. In: Kawanabe H, Mizuno N, Hosoya K (eds) Freshwater fishes of Japan. Yama-kei Publishers Co., Ltd., Tokyo, pp 382–385 (in Japanese)

Saitoh K, Katano O, Koizumi A (1988) Movement and spawning of several freshwater fishes in temporary waters around paddy fields. Japanese J Ecol 38:35–47 (in Japanese with English abstract). https://doi.org/10.18960/seitai.38.1_35

Satoh M, Sakata H, Tuan DD, Fujiki T (1998) Runoff characteristics of return flow from a paddy field area. Trans Japan Soc Irrig Drainage Reclam Eng 198:985–991 (in Japanese with English abstract). https://doi.org/10.11408/jsidre1965.1998.985

Shinzawa K, Koide S (1963) Land readjustment of arable land. Iwanamishoten Publishers Co., Ltd., Tokyo (in Japanese)

The Japanese Society of Irrigation, Drainage and Reclamation Engineering (2000) Agricultural engineering handbook basic edition. Japanese Society of Irrigation, Drainage and Reclamation Engineering, Tokyo (in Japanese)

Yata T (1979) The spawning period and eggs of *Gnathopogon elongatus elongatus*. Contrib Osaka Prefecture Freshw Fish Exp Stat 5:1–6 (in Japanese)

Chapter 12
Fish-Retreat Ditch Survey

Abstract The importance of biodiversity conservation functions in paddy ecosystems has become widely recognized. Many surveys of fish and dragonflies inhabiting paddy fields, irrigation and drainage canals have been conducted. It has been highlighted that it is important to secure flowing water in drainage canals year round. If drainage canals and paddy fields can be flooded, then they will contribute to the improvement of biodiversity conservation functions of birds and frogs. As in the above case, although there are structures similar to fish-retreat ditches and swales across Japan, there are few studies that confirm their effects. In this study, to address this issue, the abundance of fish inhabiting the fish-retreat ditch created in a paddy field near the sea in Kunitomi District, Obama City, Fukui Prefecture, was investigated. The fish-retreat ditch contained purely freshwater, amphidromous, brackish water and marine fish. The specimens collected in this ditch included 565 *Misgurnus anguillicaudatus*, 306 *Cobitis* sp. BIWAE type A, 132 *Tribolodon hakonensis*, 122 *Acanthogobius flavimanus*, 33 *Gymnogobius urotaenia* and 34 other fish. *M. anguillicaudatus* and *Cobitis* sp. BIWAE type A (purely freshwater fish) use the fish-retreat ditch mainly as a spawning ground during the irrigation season; their numbers dropped during the nonirrigation season. *T. hakonensis*, *G. urotaenia* (amphidromous fish) and *A. flavimanus* (brackish water/marine fish) used the fish-retreat ditch mostly as a growing ground. They swim upstream to the fish-retreat ditch in spring, grow from summer to autumn and remain until spawning. In addition, according to the survey that monitors the ascending and descending of a fish community in the fish-retreat ditch, the fish-retreat ditch is not only a spawning ground but also an overwintering ground for *M. anguillicaudatus* and *Cobitis* sp. BIWAE type A (purely freshwater fish). And the ditch was not only a growing ground but also an overwintering ground for *T. hakonensis*, *G. urotaenia* (amphidromous fish) and *A. flavimanus* (brackish water/marine fish). The fish-retreat ditch, which has had less environmental changes than rivers, drainage canals and oceans, has become a nursery for many fish. The biodiversity conservation function of the fish-retreat ditch was confirmed, and to increase this function, a network of multiple fish-retreat ditches should be constructed in the future.

This chapter is a revised version of Matsui (2021). Copyright 2021, https://doi.org/10.18960/hozen.2015, Accessed November 28, 2021; and Matsui (2022). Copyright 2022 Ecology and Civil Engineering Society, https://doi.org/10.3825/ece.21-00018, Accessed December 10, 2021.

© The Author(s), under exclusive license to Springer Nature Singapore Pte Ltd. 2022
A. Matsui, *Wetland Development in Paddy Fields and Disaster Management*,
https://doi.org/10.1007/978-981-19-3735-4_12

Keywords Amphidromous fish · Brackish and marine fish · Growing ground ·
Nursery · Overwintering ground · Purely freshwater fish · Spawning ground

12.1   Introduction

The importance of biodiversity conservation functions in paddy ecosystems has
become widely recognized. Many surveys of fish and dragonflies inhabiting paddy
fields, irrigation and drainage canals have been conducted. It has been highlighted
that it is important to secure flowing water in drainage canals year round (Matsui
and Satoh 2004; Matsui 2009). If drainage canals and paddy fields can be flooded,
then they will contribute to the improvement of biodiversity conservation functions
of birds and frogs (Iwabuchi 2003; Kurechi 2007, 2016; Washitani 2007). Thus,
many studies have confirmed the positive effects of flooding paddy fields in winter
to provide feeding grounds and resting places for birds and of installing paddy fishways to enable fish to ascend from the drainage canal to the paddy fields (Hata 1999;
Sato et al. 2008; Suzuki et al. 2004).

On the other hand, swales have been installed in an effort to flood part of the
paddy fields instead of flooding all the paddy fields. There have been cases where
a swale was installed in dry consolidated paddy fields (type 1) and cases where it
was installed in wet paddy fields (type 2) (Table 12.1). The type 1 swale is mainly
constructed on flat land that connects to rivers and lakes, and in some cases, a paddy
fishway is added. Habitat for aquatic animals can be provided by creating a year-
round flooded swale in a part of paddy fields. For example, this swale serves as an
evacuation site for *Misgurnus anguillicaudatus* and a spawning site for *Rana* sp.
during water fall and winter (Miyasato 2009). In Echizen City, Fukui Prefecture, a
strategy for creating a swale where *Ciconia boyciana* exist has been formulated. A
fish-retreat ditch in paddy fields that allows aquatic animals to move during drought or
snowfall has been developed. Hiyori et al. (2016, 2017) reported that the population
density of fish-retreat ditches is significantly higher than that of paddy fields and
that fish-retreat ditches provide a year-round aquatic animal habitat. In Sado City,
Niigata Prefecture, a certification system for creating a swale for *Nipponia nippon*
has been established. It is necessary to adopt the technology such as swales, paddy
fishways and biotopes that conserves organisms during winter flooding (Sado City
2021). The swale is a fish-retreat ditch in Echizen City. In Shiga Prefecture, an effort
is being implemented to restore the paddy environment so that lake fish can spawn
and grow under the fish nursery paddy field (Shiga Prefecture 2018). Here, the fish
that live in the drainage canal can ascend the paddy fishway and spawn in the paddy
field.

A typical example of the type 2 swale is 'nurume.' Nurume refers to a small
waterway created in a paddy field in a cold, mountainous region. When cold spring
water supplied from mountainous areas is drawn to paddy fields, the water is warmed
in the small waterway beside the paddy field before being placed in the paddy field
to avoid hindering the growth of rice. Since water is constantly flowing in a nurume,

Table 12.1 Types of fish-retreat ditch in Japan

Type	Paddy water management	Regional environment	Installation location	Height difference from the canal	Facility	Flooded area without cultivation	Name	References
1a	Dry paddy	Plain connected to rivers and lakes	Drainage side	Large	Paddy fishway	Installation	Fish-retreat ditch	This study
1b			Irrigation side	Small	None	Installation	Swale	Ishima et al. (2016)
1c			Entire drainage canal	Large	Paddy fishway	No installation	Weir canal	Shiga Prefecture (2018)
2	Wet paddy	Most upstream and adjacent to forest	Irrigation side	Small	None	Installation	Nurume	Echizen City (2020)

Source Reprinted from Matsui (2021). Copyright 2021

it functions as a refuge site for aquatic animals when the paddy fields are dried. It is effective in protecting and conserving aquatic animals (Echizen City 2020).

As in the above case, although there are structures similar to fish-retreat ditches and swales across Japan, there are few studies that confirm their effects. In this study, to address this issue, the abundance of fish inhabiting the fish-retreat ditch created in Kunitomi District, Obama City, Fukui Prefecture (35°29′57.9″N, 135°46′16.3″E), was investigated. This area corresponds to a type 1 swale (Table 12.1) but differs from other cases in that it is located in the lower reaches of the plain and is close to the sea. The swale described in Ishima et al. (2016) is located on the inside of a paddy field and has no paddy fishway, whereas the fish-retreat ditch in this study is located on the outside of the paddy field and has a paddy fishway. In this study, there is also a structural difference in that the fish-retreat ditch is connected to the paddy fishway from the drainage canal.

12.2 Methods

12.2.1 Survey Areas

The survey area was in a fish-retreat ditch created in Kunitomi District, Obama City, Fukui Prefecture (Fig. 12.1). The area is the last breeding ground for wild *Ciconia boyciana* in the country, and this species occasionally visits the survey area. In this area, the *Ciconia boyciana* Township Promotion Association has been established to restore *Ciconia boyciana* in this area.

The fish-retreat ditch and paddy fishway in this area were established as part of the Fukui Prefectural core irrigation facility stock management project (Kunitomi District, 2016–2017) to allow aquatic animals to move freely between drainage canals and paddy fields. The fish-retreat ditch was excavated from the original paddy field, with the sediment returned to the riverbed. A weed-proof sheet was placed on the slope on the drainage canal side, and weeding is carried out several times a year on the ridge in the fish-retreat ditch. Paddy fields are spread out around it, and irrigation and drainage canals surround it. Aquatic animals such as fish can move into the fish-retreat ditch through drainage canals and paddy fishways (Fig. 12.2). In addition, PVC pipes (VP150, $L = 1.1$ m) were installed so that aquatic animals could move between the fish-retreat ditch and the paddy field, but partition plates were installed during this survey period because of flooding inside the paddy field. Therefore, it was not possible for aquatic organisms to move between the fish-retreat ditch and the paddy field. The flowing water in the fish-retreat ditch is supplied from the nearby groundwater, and the ditch is flooded year round. The fish-retreat ditch is 44 m in length, 2 m in width and approximately 0.6 m in depth. The drainage from the fish-retreat ditch enters the drainage canal, and the Eko River and Kita River flow into Obama Bay. The area near the confluence of the Eko River and Kita River is a brackish water area. There is no height difference between the Eko River and Kita

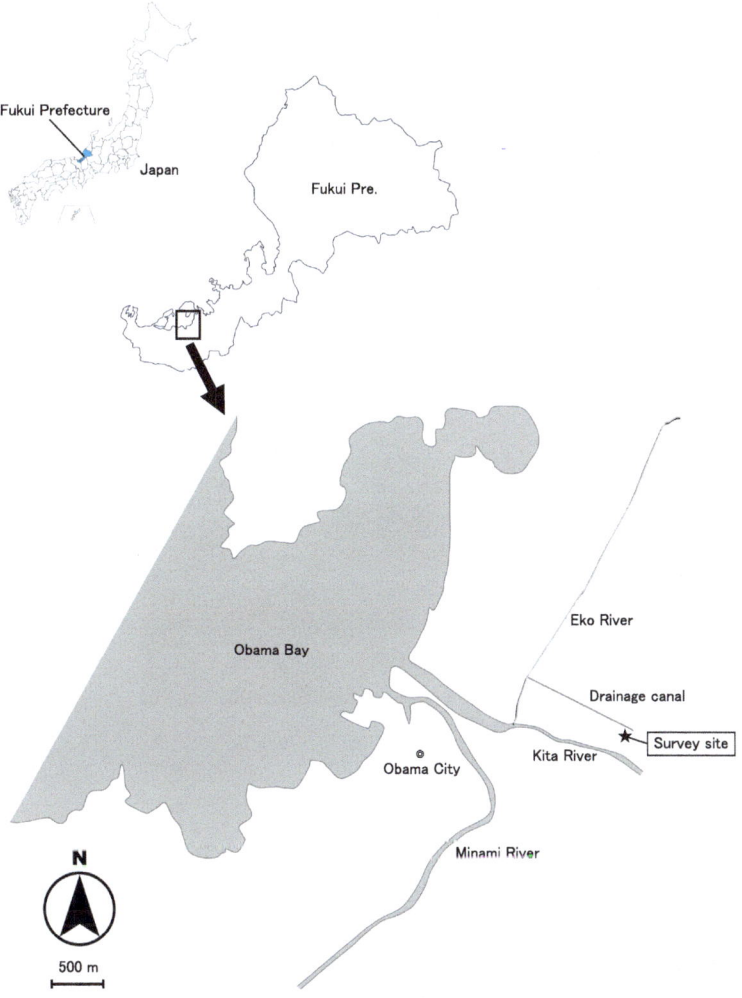

Fig. 12.1 Location of the survey sites. *Source* Reprinted from Matsui (2021, 2022). Copyright 2021, Copyright 2022 Ecology and Civil Engineering Society

River. There is usually a slight difference in elevation between the drainage canal and the Eko River (approximately 0.3 m), but they are connected when the water level rises. There is a height difference of approximately 1 m between the fish-retreat ditch and the drainage canal, but aquatic animals can move through the paddy fishway.

Figure 12.3 shows the seasonal changes in water depth in the paddy field adjacent to the fish-retreat ditch. The agricultural irrigation season is from May to August, and the nonirrigation season is from September to April. Plowing occurs in the paddy field in early April, rice planting occurs in early May, mid-season drainage

a

b

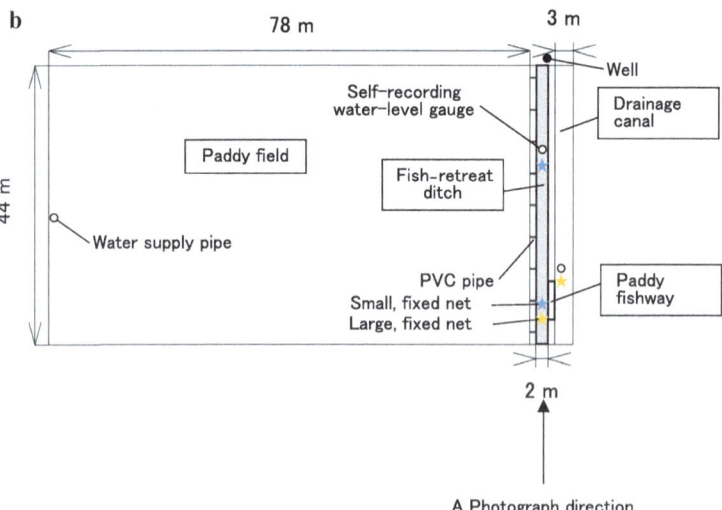

Fig. 12.2 Fish-retreat ditch: **a** photo and **b** location (*Photo* by Akira Matsui). *Notes* Photo date: June 7, 2020. A large, fixed net is installed in photo **a**. *Source* Reprinted from Matsui (2021, 2022). Copyright 2021, Copyright 2022 Ecology and Civil Engineering Society

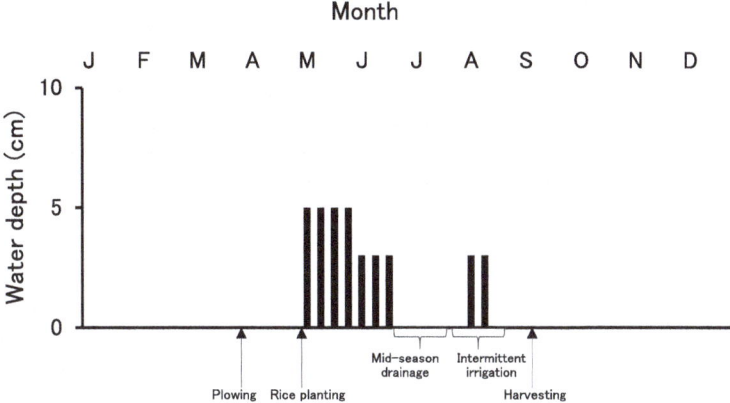

Fig. 12.3 Seasonal changes in water depth in the paddy field, Kunitomi District, Obama City, Fukui Prefecture, Japan. *Note* Referring to Nishikawa (2015). *Source* Reprinted from Matsui (2021). Copyright 2021

occurs for approximately a month in July, intermittent irrigation occurs in August, and harvesting occurs in early September.

12.2.2 Survey Methods

From May 1, 2019 to January 4, 2020, a small, fixed net (diameter of bag net 0.2 m, length of bag net 0.9 m, length of sleeve net 1 m and opening size 4 mm, Fig. 12.2b ★) was always installed in the fish-retreat ditch. The fish were collected and identified once a week. Two small, fixed nets were installed in the fish-retreat ditch, and the caught fish were placed together locally. The species were identified, and total length was measured. Then, the fish were returned to the fish-retreat ditch. To determine how each fish used the fish-retreat ditch, the seasonal prevalence of yearling fish (0-year-old fish) from the frequency distribution of the total length was determined.

From January 5 to December 27, 2020, a large, fixed net (diameter of bag net 0.4 m, length of bag net 2 m, length of sleeve net 6 m and opening size 5 mm, Fig. 12.2b ★) was always installed on the upper and lower parts of the paddy fishway that led to the fish-retreat ditch. The caught fish were collected and identified at a frequency of once a week. During the period, the position of the large, fixed net was changed to determine if the fish ascend to the fish-retreat ditch or descend from the fish-retreat ditch. The large, fixed net was set up upstream of the paddy fishway from March 1 to August 30, 2020, to confirm that the fish ascended to the fish-retreat ditch. In addition, the large, fixed net was set up downstream of the paddy fishway from January 5 to February 29, 2020, and from August 31 to December 27, 2020, to confirm that the fish descended from the fish-retreat ditch.

At the same time, a self-recording water-level gauge (manufactured by OYO Corporation, S & DL mini) was installed in the fish-retreat ditch and the drainage canal, and the water level was observed continuously every hour. Water temperature, dissolved oxygen (DO) concentration, hydrogen ion concentration (pH) and electrical conductivity (EC) were measured using a portable multi-item water quality meter (manufactured by DKK-TOA Corporation, WQC-24) during each survey.

12.3   Results

12.3.1   Aquatic Animal Biomass in the Fish-Retreat Ditch

As a result of the field survey, 12 species and 1192 individuals of fish and shellfish were collected (Table 12.2): 565 *Misgurnus anguillicaudatus* (47.4%), 306 *Cobitis* sp. BIWAE type A (25.7%), 132 *Tribolodon hakonensis* (11.1%), 122 *Acanthogobius flavimanus* (10.2%), 33 *Gymnogobius urotaenia* (2.8%) and 34 other fish (2.8%). The life-history fish types were 886 purely freshwater fish (74.3%), 184 amphidromous and catadromous fish (15.4%) and 122 brackish and marine fish (10.2%) (Fig. 12.4).

Table 12.2   Fish and shellfish caught in the fish-retreat ditch

Species	Life type	Number of individuals	Ratio (%)
Tanakia limbata	Pure freshwater type	1	0.1
Opsariichthys platypus	Pure freshwater type	1	0.1
Phoxinus lagowskii steindachneri	Pure freshwater type	12	1.0
Tribolodon hakonensis	Amphidromous type	132	11.1
Gnathopogon elongatus elongatus	Pure freshwater type	1	0.1
Misgurnus anguillicaudatus	Pure freshwater type	565	47.4
Cobitis sp. BIWAE type A	Pure freshwater type	306	25.7
Gymnogobius petschiliensis	Amphidromous type	3	0.3
Gymnogobius urotaenia	Amphidromous type	33	2.8
Acanthogobius flavimanus	Brackish and marine type	122	10.2
Rhinogobius nagoyae	Amphidromous type	3	0.3
Eriocheir japonicus	Catadromous type	13	1.1
Total		1192	100

Note Two small, fixed nets were used in the fish-retreat ditch
Source Reprinted from Matsui (2021). Copyright 2021

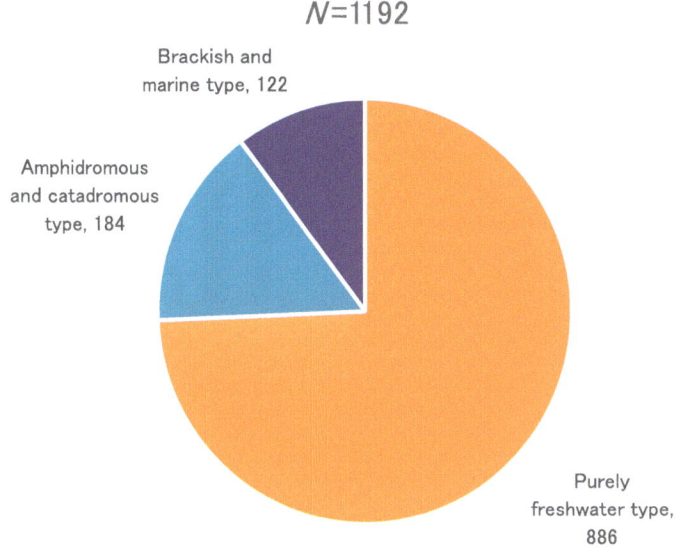

Fig. 12.4 Fish and shellfish caught in the fish-retreat ditch (by life-history type). *Source* Reprinted from Matsui (2021). Copyright 2021

Figure 12.5 shows the seasonal changes in the capture number of *Misgurnus anguillicaudatus*, *Cobitis* sp. BIWAE type A, *Tribolodon hakonensis*, *Acanthogobius flavimanus* and *Gymnogobius urotaenia*, which were abundant. *Misgurnus anguillicaudatus* was captured from May to November 2019. Its abundance in May–August showed more than 100 individuals/month, but that in September–November decreased significantly (Fig. 12.5a). *Cobitis* sp. BIWAE type A was caught from May to October 2019. The abundance in May–August was less than that of *Misgurnus anguillicaudatus* but was 60–90 individuals/month. However, the number of *Cobitis* sp. BIWAE type A caught in September–October dropped sharply (Fig. 12.5b). *Tribolodon hakonensis* was captured from August to December 2019. Its abundance in August–September was low, but that in October–December increased significantly (Fig. 12.5c). *Acanthogobius flavimanus* was captured from June to December 2019, and its abundance was highest in August and gradually decreased after September (Fig. 12.5d). *Gymnogobius urotaenia* was caught from July to December 2019, and its abundance was highest in August (Fig. 12.5e).

Figure 12.6 shows the frequency distribution of *Misgurnus anguillicaudatus*, *Cobitis* sp. BIWAE type A, *Tribolodon hakonensis*, *Acanthogobius flavimanus* and *Gymnogobius urotaenia* by total length. The smallest *Misgurnus anguillicaudatus*, 4–6 cm in size, appeared in June 2019 (Fig. 12.6a). The smallest *Cobitis* sp. BIWAE type A, 4–6 cm, was captured from May to August 2019 (Fig. 12.6b). The smallest *Tribolodon hakonensis*, 4–6 cm, was confirmed from October to December 2019 (Fig. 12.6c). The smallest *Acanthogobius flavimanus*, 6–8 cm, appeared from July to September 2019, and its frequency distribution of total length tended to increase until

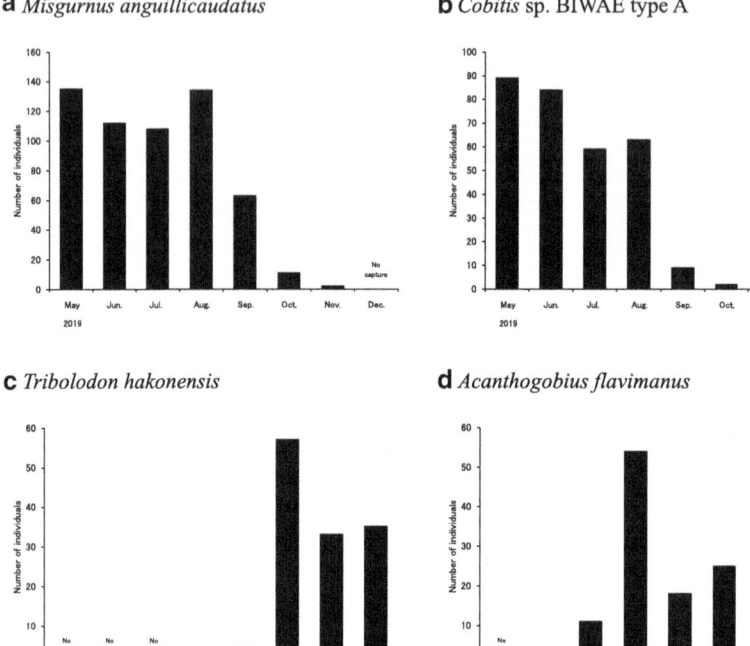

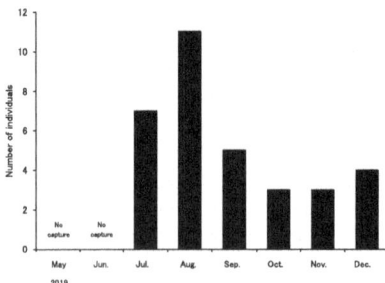

Fig. 12.5 Seasonal changes in the capture number of **a** *Misgurnus anguillicaudatus*, **b** *Cobitis* sp. BIWAE type A, **c** *Tribolodon hakonensis*, **d** *Acanthogobius flavimanus* and **e** *Gymnogobius urotaenia* in the fish-retreat ditch. *Source* Reprinted from Matsui (2021). Copyright 2021

December (Fig. 12.6d). The smallest *Gymnogobius urotaenia,* 2–4 cm, was collected in July 2019, and its frequency distribution of total length tended to increase from August to December (Fig. 12.6e).

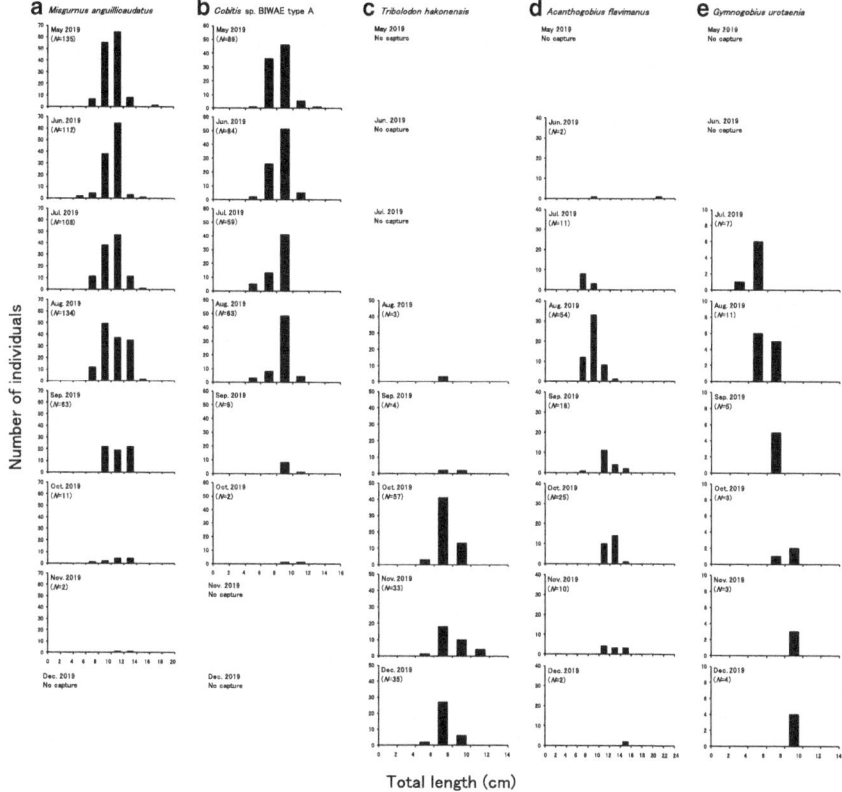

Fig. 12.6 Frequency distribution of **a** *Misgurnus anguillicaudatus*, **b** *Cobitis* sp. BIWAE type A, **c** *Tribolodon hakonensis*, **d** *Acanthogobius flavimanus* and **e** *Gymnogobius urotaenia* caught in the fish-retreat ditch by total length. *Source* Reprinted from Matsui (2021). Copyright 2021

12.3.2 Aquatic Animal Biomass Ascending to the Fish-Retreat Ditch

As a result of the field survey, 13 species and 708 aquatic animals ascended to the fish-retreat ditch (Table 12.3): 244 *Misgurnus anguillicaudatus* (34.5%), 209 *Tribolodon hakonensis* (29.5%), 79 *Cobitis* sp. BIWAE type A (11.2%), 46 *Gymnogobius urotaenia* (6.5%), 40 *Rhinogobius nagoyae* (5.6%), 29 *Opsariichthys platypus* (4.1%) and 61 other aquatic animals (8.6%). The life-history type was 389 purely freshwater fish (54.9%), 313 amphidromous and catadromous fish (44.2%) and 6 brackish and marine fish (0.9%) (Fig. 12.7).

Figure 12.8 shows the seasonal changes in the capture number of *Misgurnus anguillicaudatus*, *Cobitis* sp. BIWAE type A, *Tribolodon hakonensis*, *Acanthogobius flavimanus* and *Gymnogobius urotaenia*, which were abundant in the fish-retreat ditch. *Misgurnus anguillicaudatus* was captured from March to August 2020, and

Table 12.3 Aquatic animals ascending to the fish-retreat ditch

Species	Life type	Number of individuals	Ratio (%)
Opsariichthys platypus	Pure freshwater type	29	4.1
Phoxinus lagowskii steindachneri	Pure freshwater type	6	0.8
Tribolodon hakonensis	Amphidromous type	209	29.5
Gnathopogon elongatus elongatus	Pure freshwater type	16	2.3
Misgurnus anguillicaudatus	Pure freshwater type	244	34.5
Cobitis sp. BIWAE type A	Pure freshwater type	79	11.2
Silurus asotus	Pure freshwater type	10	1.4
Odontobutis obscura	Pure freshwater type	4	0.6
Gymnogobius urotaenia	Amphidromous type	46	6.5
Acanthogobius flavimanus	Brackish and marine type	6	0.8
Rhinogobius nagoyae	Amphidromous type	40	5.6
Eriocheir japonicus	Catadromous type	18	2.5
Mauremys japonica	Pure freshwater type	1	0.1
Total		708	100

Note A large, fixed net was used on the upper part of the paddy fishway leading to the fish-retreat ditch

Source Reprinted from Matsui (2022). Copyright 2022 Ecology and Civil Engineering Society

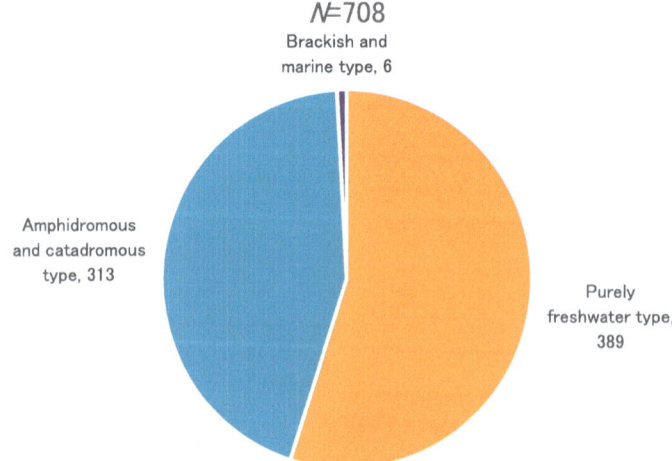

Fig. 12.7 Aquatic animals ascending to the fish-retreat ditch (by life-history type). *Source* Reprinted from Matsui (2022). Copyright 2022 Ecology and Civil Engineering Society

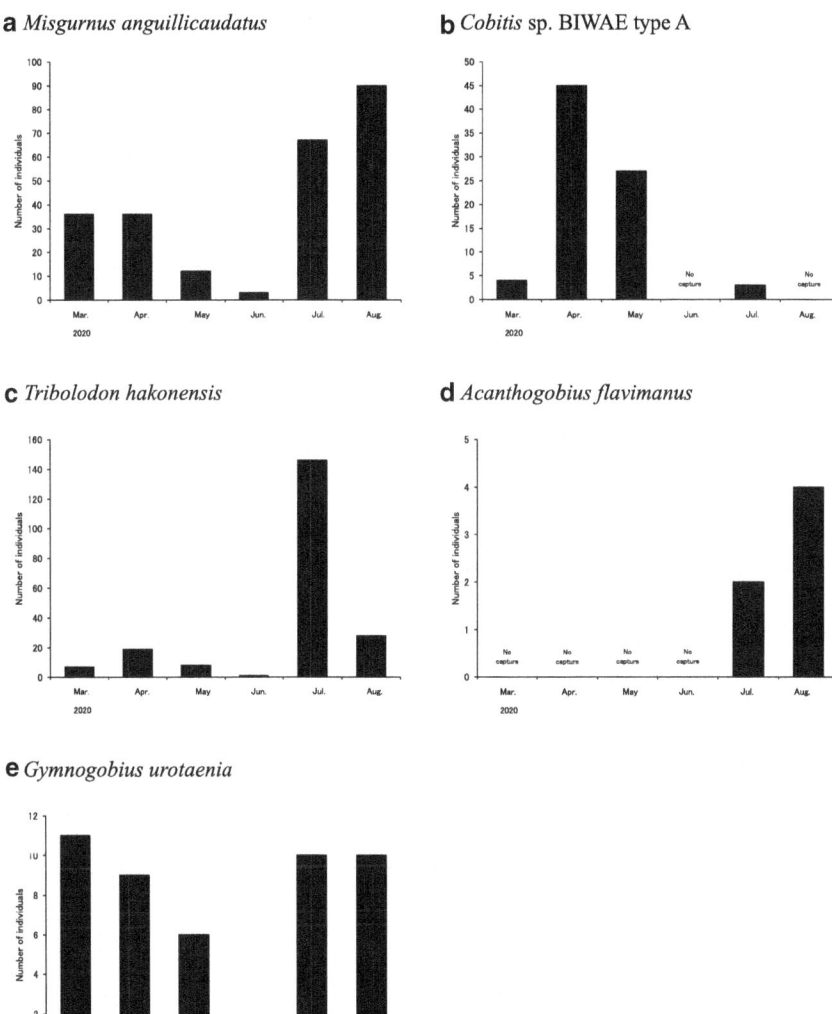

Fig. 12.8 Seasonal changes in capture number of **a** *Misgurnus anguillicaudatus*, **b** *Cobitis* sp. BIWAE type A, **c** *Tribolodon hakonensis*, **d** *Acanthogobius flavimanus* and **e** *Gymnogobius urotaenia* ascending to the fish-retreat ditch. *Source* Reprinted from Matsui (2022). Copyright 2022 Ecology and Civil Engineering Society

the number of its ascension events in March–April and July–August was relatively high (Fig. 12.8a). *Cobitis* sp. BIWAE type A was caught in March–May and July 2020, and the number of its ascension events from April to May was significantly higher than that in other months (Fig. 12.8b. *Tribolodon hakonensis* was captured from March to August 2020, and the number of its ascension events in July was significantly higher than that in other months (Fig. 12.8c). *Acanthogobius flavimanus* was captured in July–August 2020, and the number of its ascension events was not confirmed from March to June (Fig. 12.8d). *Gymnogobius urotaenia* was caught in March–May and July–August 2020, and the number of its ascension events was not confirmed in June (Fig. 12.8e).

Figure 12.9 shows the frequency distribution of *Misgurnus anguillicaudatus, Cobitis* sp. BIWAE type A, *Tribolodon hakonensis, Acanthogobius flavimanus* and *Gymnogobius urotaenia* by total length. 0-year-old *Misgurnus anguillicaudatus* ascended at a size of 4–6 cm from June to August 2020. From March to May and from July to August, fish that were 1-year-old and older fish (including 0-year-old fish) ascended (Fig. 12.9a). *Cobitis* sp. BIWAE type A ascended at 6 cm or more when they were 1-year-old and older fish in March–May and July 2020. During the survey period, 0-year-old fish that were 4–6 cm did not ascend (Fig. 12.9b). 0-year-old *Tribolodon hakonensis* ascended at 4–6 cm in April and June–August 2020. In

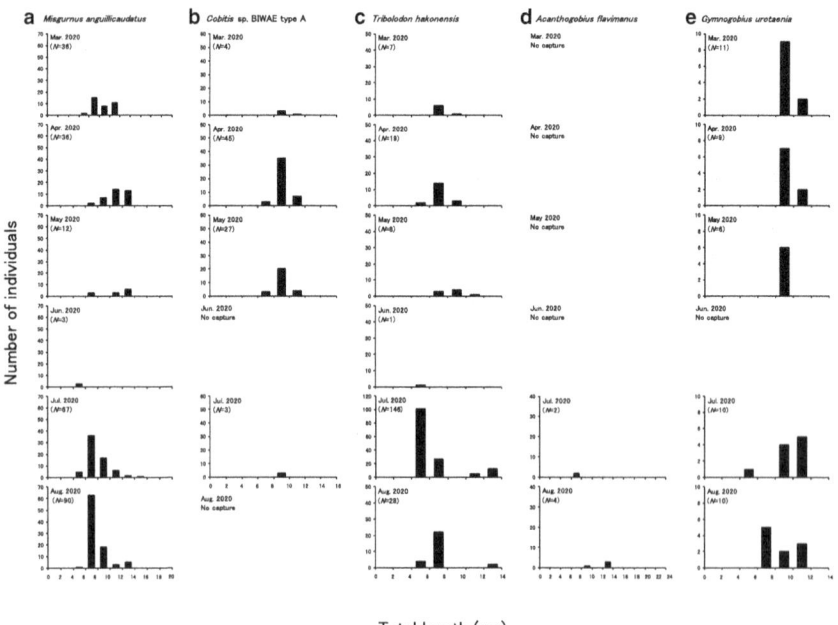

Total length (cm)

Fig. 12.9 Frequency distribution of **a** *Misgurnus anguillicaudatus*, **b** *Cobitis* sp. BIWAE type A, **c** *Tribolodon hakonensis*, **d** *Acanthogobius flavimanus* and **e** *Gymnogobius urotaenia* ascending to the fish-retreat ditch by total length. *Source* Reprinted from Matsui (2022). Copyright 2022 Ecology and Civil Engineering Society

March–May and July–August, 1-year-old and older fish (including 0-year-old fish) increased in size to 6 cm or more (Fig. 12.9c). 0-year-old *Acanthogobius flavimanus* ascended at 6–8 cm in July 2020. In August, 1-year-old and older fish at 8 cm or more ascended (Fig. 12.9d). 0-year-old *Gymnogobius urotaenia* ascended at 4–6 cm in July 2020. In March–May and July–August, 1-year-old and older fish (including 0-year-old fish) increased in size to 6 cm or more (Fig. 12.9e).

12.3.3 Aquatic Animal Biomass Descending from the Fish-Retreat Ditch

As a result of the field survey, 6 species and 103 aquatic animals descended from the fish-retreat ditch (Table 12.4): 88 *Misgurnus anguillicaudatus* (85.4%), 7 *Gymnogobius urotaenia* (6.8%), 3 *Tribolodon hakonensis* (2.9%), 2 *Phoxinus lagowskii steindachneri*, 2 *Acanthogobius flavimanus* (each 1.9%) and 1 *Opsariichthys platypus* (1.0%). The life-history types were 91 purely freshwater fish (88.4%), 10 amphidromous fish (9.7%) and 2 brackish and marine fish (1.9%) (Fig. 12.10).

Figure 12.11 shows the seasonal changes in the capture number of *Misgurnus anguillicaudatus*, *Cobitis* sp. BIWAE type A, *Tribolodon hakonensis*, *Acanthogobius flavimanus* and *Gymnogobius urotaenia*, which were abundant in the fish-retreat ditch. *Misgurnus anguillicaudatus* was caught in January–February and September–December 2020, and the number of its descents in September was significantly high (Fig. 12.11a). *Cobitis* sp. BIWAE type A was not caught at all (Fig. 12.11b).

Table 12.4 Aquatic animals descending from the fish-retreat ditch

Species	Life type	Number of individuals	Ratio (%)
Opsariichthys platypus	Pure freshwater type	1	1.0
Phoxinus lagowskii steindachneri	Pure freshwater type	2	1.9
Tribolodon hakonensis	Amphidromous type	3	2.9
Misgurnus anguillicaudatus	Pure freshwater type	88	85.4
Gymnogobius urotaenia	Amphidromous type	7	6.8
Acanthogobius flavimanus	Brackish and marine type	2	1.9
Total		103	100

Note A large, fixed net was used on the lower part of the paddy fishway leading to the fish-retreat ditch

Source Reprinted from Matsui (2022). Copyright 2022 Ecology and Civil Engineering Society

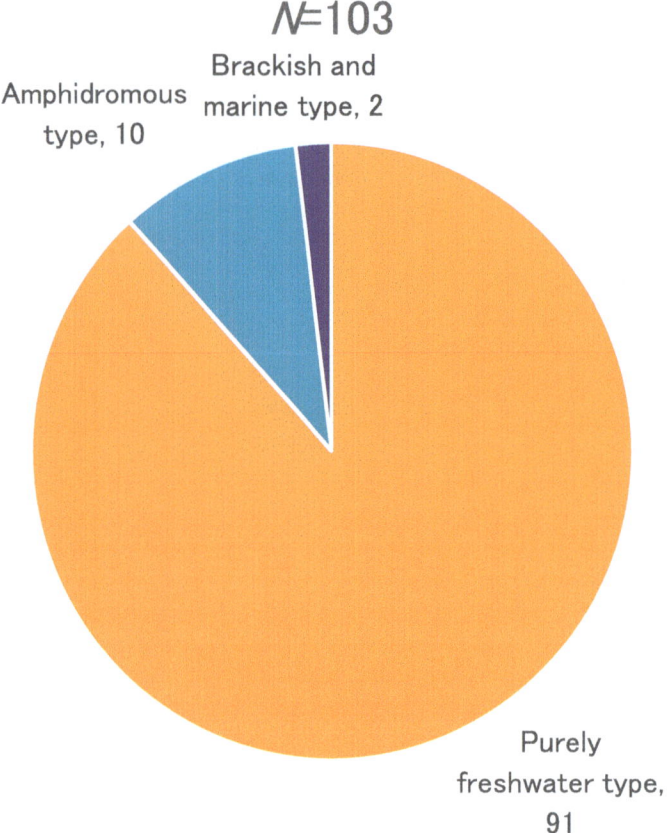

Fig. 12.10 Aquatic animals descending from the fish-retreat ditch (by life-history type). *Source* Reprinted from Matsui (2022). Copyright 2022 Ecology and Civil Engineering Society

Tribolodon hakonensis was captured from October to December 2020, and the number of its descents was generally low (Fig. 12.11c). *Acanthogobius flavimanus* was captured in September and December 2020, and the number of its descents was generally low (Fig. 12.11d). *Gymnogobius urotaenia* was caught in January and October 2020, and the number of its descents in January was relatively high (Fig. 12.11e).

Figure 12.12 shows the frequency distribution of *Misgurnus anguillicaudatus*, *Cobitis* sp. BIWAE type A, *Tribolodon hakonensis*, *Acanthogobius flavimanus* and *Gymnogobius urotaenia* by total length. 0-year-old *Misgurnus anguillicaudatus* descended when they were 4–6 cm from September to December 2020. From January to February and from September to December, 1-year-old and older fish (including 0-year-old fish) descended when they were 6 cm or more (Fig. 12.12a). *Cobitis* sp. BIWAE type A did not descend during the survey (Fig. 12.12b). 1-year-old *Tribolodon hakonensis* descended when they were 6–8 cm (including 0-year-old

a *Misgurnus anguillicaudatus*

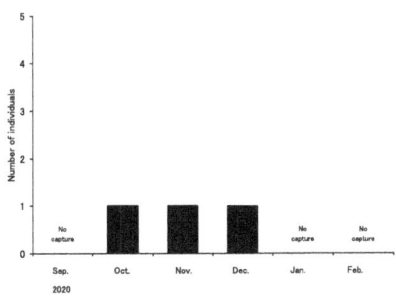

b *Cobitis* sp. BIWAE type A

No capture

c *Tribolodon hakonensis*

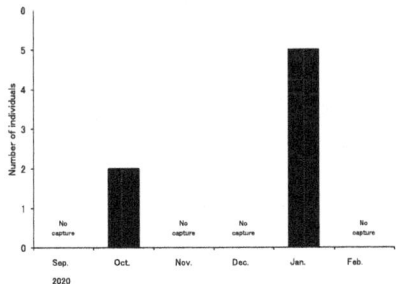

d *Acanthogobius flavimanus*

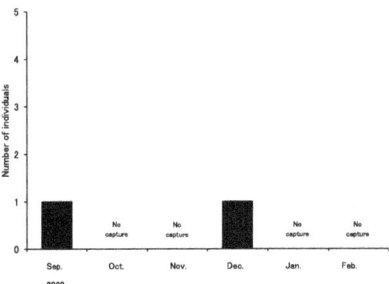

e *Gymnogobius urotaenia*

Fig. 12.11 Seasonal changes in the capture number of **a** *Misgurnus anguillicaudatus*, **b** *Cobitis* sp. BIWAE type A, **c** *Tribolodon hakonensis*, **d** *Acanthogobius flavimanus* and **e** *Gymnogobius urotaenia* descending from the fish-retreat ditch. *Source* Reprinted from Matsui (2022). Copyright 2022 Ecology and Civil Engineering Society

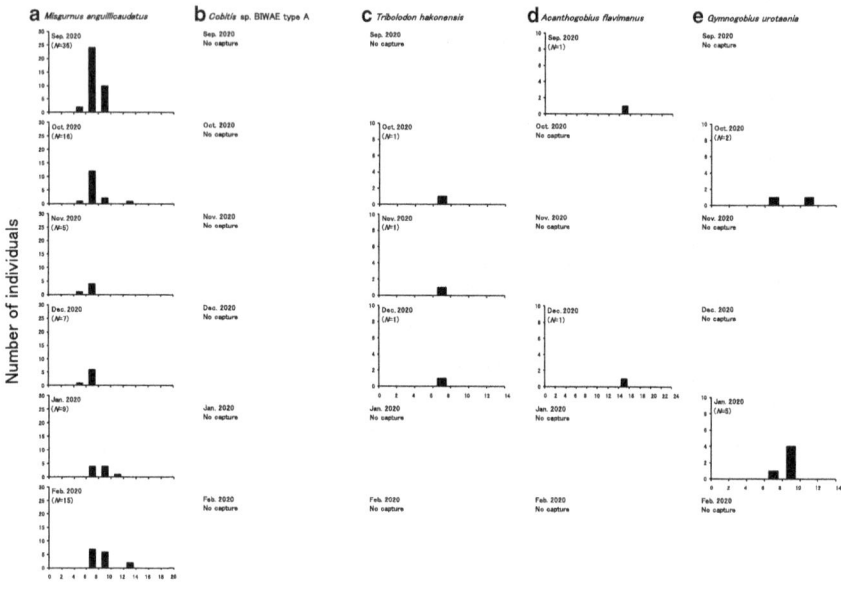

Total length (cm)

Fig. 12.12 Frequency distribution of **a** *Misgurnus anguillicaudatus,* **b** *Cobitis* sp. BIWAE type A, **c** *Tribolodon hakonensis,* **d** *Acanthogobius flavimanus* and **e** *Gymnogobius urotaenia* descending from the fish-retreat ditch by total length. *Source* Reprinted from Matsui (2022). Copyright 2022 Ecology and Civil Engineering Society

fish) in October–December 2020 (Fig. 12.12c). 1-year-old and older *Acanthogobius flavimanus* descended when they were 14–16 cm in September and December 2020 (Fig. 12.12d). 1-year-old and older *Gymnogobius urotaenia* descended when they were 6 cm or more (including 0-year-old fish) in January and October 2020 (Fig. 12.12e).

12.3.4   Water Depth, Water Temperature and Water Quality in the Fish-Retreat Ditch

Figure 12.13 shows the seasonal changes in precipitation and water depth measured by a self-recording water-level gauge installed in the fish-retreat ditch and drainage canal. The maximum water depth in the fish-retreat ditch was 56.7 cm, the minimum water depth was 7.6 cm, and the average water depth was 34.3 ($\pm$7.0) cm; thus, it was flooded year round. On the other hand, the maximum water depth in the drainage canal was 71.5 cm, the minimum water depth was 0.0 cm, and the average water depth was 20.7 ($\pm$7.7) cm; thus, it was not flooded year round.

Table 12.5 shows the results of water temperature and water quality in the fish-

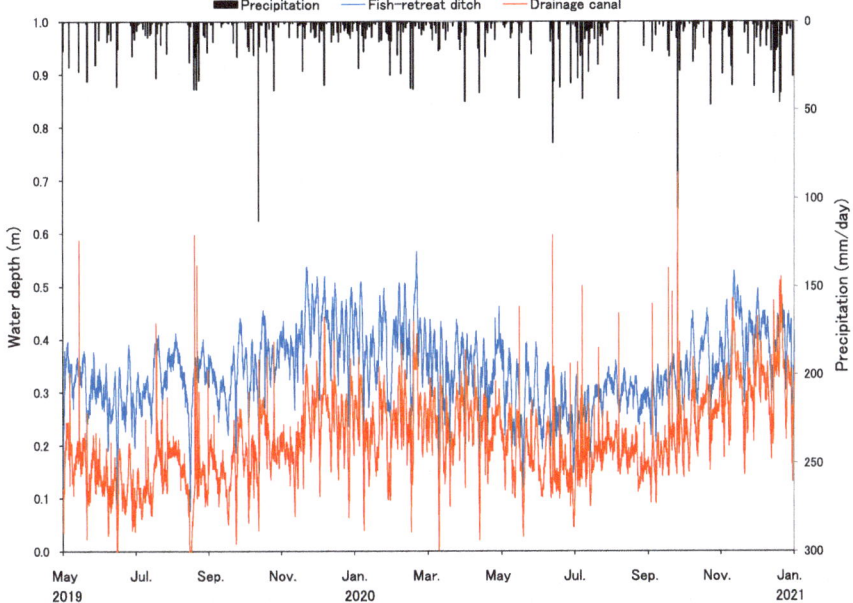

Fig. 12.13 Seasonal changes in precipitation and water depth measured by a self-recording water-level gauge in the fish-retreat ditch and drainage canal. *Source* Reprinted from Matsui (2021, 2022). Copyright 2021, Copyright 2022 Ecology and Civil Engineering Society

Table 12.5 Water temperature and water quality in the fish-retreat ditch

Survey item	Maximum	Minimum	Average ± Standard deviation
Water temperature (°C)	32.8	4.9	16.1 ± 3.7
DO (mg/L)	18.8	5.7	8.8 ± 1.6
pH	9.5	6.3	7.1 ± 0.5
EC (mS/m)	67.2	7.7	13.6 ± 9.1

Source Reprinted from Matsui (2021, 2022). Copyright 2021, Copyright 2022 Ecology and Civil Engineering Society

retreat ditch. The maximum water temperature was 32.8 °C, the minimum water temperature was 4.9 °C, and the average water temperature was 16.1 (±3.7) °C. The water temperature rose in the summer, but it did not drop substantially in the winter because it was groundwater. The maximum DO value was 18.8 mg/L, the minimum DO value was 5.7 mg/L, and the average DO value was 8.8 (±1.6) mg/L. The maximum pH value was 9.5, the minimum pH value was 6.3, and the average pH value was 7.1 (±0.5). The maximum EC value was 67.2 mS/m, the minimum EC value was 7.7 mS/m, and the average EC value was 13.6 (±9.1) mS/m.

12.4 Discussion

12.4.1 Characteristics of the Major Fish Species

(1) *Misgurnus anguillicaudatus*

Regarding the characteristics of *Misgurnus anguillicaudatus*, Miyaji et al. (1963) reported that this species reaches a total length of 2 cm at 10 days, a total length of 8 to 10 cm at 1 year and a total length of 10 to 12 cm at 2 years. Kubota (1961) reported that this species reaches a total length of 69 mm 3 months after hatching and 72–73 mm 9 months after hatching. 12 months after hatching, females reach 104 mm, and males reach 97 mm; 13–14 months after hatching, females reach 106 mm, and males reach 99 mm; and 22–23 months after hatching, females reach 121 mm, and males reach 108 mm. From the above information, in this area, the smallest *Misgurnus anguillicaudatus* individuals that appeared in the fish-retreat ditch in June 2019 were 4–6 cm and presumed to be yearling fish (0-year-old fish).

The number of individuals of this species decreased in September–November 2019, and the number of 0-year-old fish and 1-year-old or older fish fell in January–February and September–December 2020. The reason for the decrease in the number of fish caught in September–November 2019 was that the fish descended from the fish-retreat ditch. However, the total number of individuals ascending to the fish-retreat ditch was 244, but the total number of individuals descending from the fish-retreat ditch was 88. Therefore, all the individuals who lived in the fish-retreat ditch did not descend. It is estimated that some (156 individuals in this survey) dove into the mud in the fish-retreat ditch and overwintered.

This species was caught in the fish-retreat ditch from May to November 2019, and the 0-year-old fish appeared in June 2019. 1-year-old or older fish ascended in March–May 2020, and 0-year-old fish ascended in June–August 2020. Therefore, the 0-year-old fish that appeared in June 2019 were from adult fish (1-year-old or older fish) that ascended to the fish-retreat ditch and spawned and hatched, adult fish (1-year-old or older fish) that overwintered in the fish-retreat ditch from the previous year and spawned and hatched and 0-year-old fish that ascended to the fish-retreat ditch. It is estimated that 156 adult fish overwintered in the fish-retreat ditch, 84 adult fish ascended to the fish-retreat ditch between March and May 2020, and three 0-year-old fish ascended to the fish-retreat ditch in June 2020. The proportion of the 0-year-old fish that spawned and hatched in the drainage canals and rivers was extremely small. It is estimated that the adult fish that overwintered in the fish-retreat ditch from the previous year were involved in spawning at approximately twice the rate of the adult fish that ascended to the fish-retreat ditch the current year.

(2) *Cobitis* sp. BIWAE type A

Regarding the characteristics of *Cobitis* sp. BIWAE type A, Nakajima and Uchiyama (2017) reported that this species is expected to live for more than two years outdoors, and its total length reaches 90–130 mm. Considering the abovementioned life history of *Misgurnus anguillicaudatus*, it is estimated that the smallest individual of 4–6 cm that appeared from May to August 2019 in this area was a 0-year-old fish.

The number of this species decreased in September–October 2019. There was no descent in September–December 2020, so the decrease in the number of individuals in September–October 2019 was presumed to occur because most of the individuals who lived in the fish-retreat ditch were diving into the mud and overwintering. Nakajima and Uchiyama (2017) reported that this species inhabits the middle of river basins and lives only in permanent waters, and it is thought that this species also overwinters in the permanent waters of the fish-retreat ditch. However, it was found that it inhabited not only the middle of river basins but also the fish-retreat ditch created in the paddy irrigation and drainage system.

This species was caught in the fish-retreat ditch in May–October 2019, and the 0-year-old fish appeared in May–August 2019. Since the 0-year-old fish did not ascend to the fish-retreat ditch in 2020, the 0-year-old fish that appeared in May–August 2019 were from adult fishes (1-year-old or older fish) that ascended to the fish-retreat ditch and spawned and hatched and adult fishes (1-year-old or older fish) that overwintered in the fish-retreat ditch the previous year and spawned and hatched. While 79 adult fish were estimated to have overwintered in the fish-retreat ditch, 76 adult fish ascended to the fish-retreat ditch from March to May 2020. Thus, it is estimated that the 0-year-old fish that appeared in the fish-retreat ditch were involved in spawning at almost the same rate as the adult fish that overwintered in the fish-retreat ditch the previous year and the adult fish that ascended to the fish-retreat ditch the current year.

(3) *Tribolodon hakonensis*

Regarding the characteristics of *Tribolodon hakonensis*, Sakai (1989) reported that the spawning season of this species is from spring to summer. This species grows into a juvenile fish that is 2–3 cm in 20–30 days and grows to a size of 5–10 cm in a year. According to this study, the smallest individual caught from October to December 2019 was 4–6 cm and was a 0-year-old fish, and 6–10 cm individual that appeared from August to September 2019 was a 1-year-old fish.

This species was caught in the fish-retreat ditch from August to December 2019, and the 0-year-old fish appeared from October to December 2019. As the 0-year-old fish ascended in April and June–August 2020, the 0-year-old fish that appeared from October to December 2019 were from those that spawned and hatched in the rivers and drainage canals. In particular, the number of 0-year-old fish that ascended to the fish-retreat ditch significantly increased in July 2020 because it is estimated that the amount of activity increased as the water temperature rose. On the other hand, while the total number of fish ascending to the fish-retreat ditch was 209, the number of those descending decreased significantly to 3 individuals. Therefore, it is suggested that the 0-year-old fish may ascend to and overwinter in the fish-retreat ditch.

(4) *Acanthogobius flavimanus*

Regarding the characteristics of *Acanthogobius flavimanus*, Nakamura (2002) reported that juvenile fish whose total length was 20–40 mm in May, 40–70 mm in September and 40–120 mm in October were caught in Hinuma, Ibaraki Prefecture. Sakai et al. (2000) reported that these species spawn from mid-March to late May, and the 0-year-old fish grow rapidly and reach approximately 100 mm in early November in Matsushima Bay, Miyagi Prefecture. Tsuji (1989) reported that the spawning season of this species is from January to March in Kyushu and southwestern Shikoku, from February to May in Tokyo and from March to May in Sendai City, Miyagi Prefecture. Many immature fish ascend the lower reaches of the river in summer. From the above information, in this study, the smallest individual that was 6–8 cm and appeared from July to September 2019 in this area was presumed to be a 0-year-old fish.

This species was caught in the fish-retreat ditch from June to December 2019, and 0-year-old fish appeared from July to September 2019. The 0-year-old fish ascended in July 2020, which coincided with the time when the 0-year-old fish appeared from July to September 2019. Since they did not ascend from March to June 2020, it is estimated that this species spawned and hatched in Obama Bay during this period. While 122 individuals were collected in the fish-retreat ditch, only 6 individuals ascended, and 2 individuals descended. Thus, the 0-year-old fish may not only ascend to but also overwinter in the fish-retreat ditch.

(5)   *Gymnogobius urotaenia*

Regarding the characteristics of *Gymnogobius urotaenia*, Dotsu (1955) reported that this species occurs in the suburbs of Fukuoka City and reaches a total length of 55–73 mm in the first year of its life and a maximum total length of 102 mm in the second year of its life; it also has a lifespan of more than three years. The spawning season of this species is from January to May near Fukuoka City (Dotsu 1955). The spawning season of this species begins in mid-May when the water temperature exceeds 15 °C and continues to late June in southern Hokkaido (Ishino 1989). From the above information, the smallest individual that was 2–4 cm and appeared in July 2019 in this area was presumed to be a 0-year-old fish.

This species was caught in the fish-retreat ditch from July to December 2019, and 0-year-old fish appeared in July 2019. The 0-year-old fish ascended in July 2020, which coincided with the time when the 0-year-old fish appeared in July 2019. Since it did not ascend in June 2020, this species likely spawned and hatched in the rivers and drainage canals during this period. On the other hand, the total number of fish ascending to the fish-retreat ditch was 46, while the total number of fish descending decreased to 7 individuals. Therefore, the 0-year-old fish may not only ascend to but also overwinter in the fish-retreat ditch.

12.4.2   Fish Types Occurring in the Fish-Retreat Ditch

From the results of this survey, the fish using the fish-retreat ditches can be classified into two types. *Misgurnus anguillicaudatus* and *Cobitis* sp. BIWAE type A (purely freshwater fish) mainly use the fish-retreat ditch as a spawning ground. They spawn in the fish-retreat ditch during the irrigation season and descend during the nonirrigation season. *Tribolodon hakonensis*, *Gymnogobius urotaenia* (amphidromous fish) and *Acanthogobius flavimanus* (brackish water/marine fish) mostly use the fish-retreat ditch as a growing ground. They ascend to the fish-retreat ditch from rivers, drainage canals and oceans in spring, grow from summer to autumn and remain in the fish-retreat ditch up to spawning. The difference between the two types is whether they lay eggs in the fish-retreat ditch, and the fish-retreat ditch appears to function as a habitat for 0-year-old fish and 1-year-old or older fish of the two types.

Regarding *Misgurnus anguillicaudatus* and *Cobitis* sp. BIWAE type A, which use the fish-retreat ditch as a spawning ground, approximately 60% of the formed ascended, and all of the latter ascended and remained and overwintered in the fish-retreat ditch. As *Tribolodon hakonensis*, *Gymnogobius urotaenia* and *Acanthogobius flavimanus* use the fish-retreat ditch as a growing ground, they may also remain and overwinter in the fish-retreat ditch.

Based on the above information, it was determined that the fish-retreat ditch is an important place for spawning, growing and overwintering for not only purely freshwater fish but also amphidromous and brackish water/marine fish. Fish-retreat ditches, which experience fewer environmental changes than rivers, drainage canals and oceans are thought to be nurseries for many fish.

12.4.3 Functional Significance of a Fish-Retreat Ditch

This study revealed that not only purely freshwater fish but also amphidromous fish and brackish water/marine fish live in the fish-retreat ditch in areas near the sea, such as those in Kunitomi District in Obama City, Fukui Prefecture, Japan. If these fish ascend from the drainage canal through the paddy fishway to the paddy field where there is no fish-retreat ditch, then it is possible that they spawn and grow in the paddy field. However, in this case, according to the water management of the paddy rice cultivation area, it will be dried for approximately one month in July. Therefore, a refuge site for these fish is required. That is, the fish-retreat ditch should function as this refuge site. Matsui (2009) estimates that *Misgurnus anguillicaudatus* spawn and hatch mainly in lateral and farm drains during the irrigation season and descend downstream of main drains during the nonirrigation season to overwinter. In the results of this survey, *Misgurnus anguillicaudatus* and *Cobitis* sp. BIWAE type A mainly used the fish-retreat ditch during the irrigation season and were not caught after that. Both species overwinter in the fish-retreat ditch provides important information when considering the maintenance of the fish-retreat ditch in the future. For example, if wintering is carried out in the mud even during the nonirrigation season, the mud should be maintained at the minimum necessary amount. This action is needed not only for *Misgurnus anguillicaudatus* and *Cobitis* sp. BIWAE type A but also *Tribolodon hakonensis*, *Gymnogobius urotaenia* and *Acanthogobius flavimanus*.

12.4.4   Management Issues Related to the Fish-Retreat Ditch

In this area, because partition plates were installed, aquatic animals could not move between the fish-retreat ditch and the paddy field, even with PVC pipes installed, because the water level of the paddy field was higher than that of the fish-retreat ditch; in addition, the water flowed from the paddy field through the PVC pipes to the fish-retreat ditch; and thus, the paddy field could not be flooded. On the other hand, when the paddy field is dried during mid-season drainage, the water level of the fish-retreat ditch needs to be lower than that of the paddy field. To satisfy the above two requirements, it is important to design the drainage outlet of a fish-retreat ditch so that the farmer can adjust the water level.

12.4.5   Future Outlook

Currently, there is only one fish-retreat ditch in this area, but it is expected that its biodiversity conservation function will be enhanced by establishing multiple locations and networks of fish-retreat ditches. Once this occurs, multiple fish-retreat ditches should be established at drainage canals along the Eko River that flow into Obama Bay because purely freshwater fish, amphidromous fish and brackish water/marine fish ascend from Obama Bay to the fish-retreat ditch through the Eko River. This scenario enables the fish to move shorter distances. Considering that the maximum movement distance of *Misgurnus anguillicaudatus* that uses temporary waters for spawning is 500 m (Nishida et al. 2006), an interval of approximately 500 m is proposed.

If the fish-retreat ditch and the paddy field can be flooded, then this area can be conserved and contribute greatly to biodiversity conservation. However, the issues related to flooding paddy fields in winter are as follows: (1) It is difficult to secure water in winter. The larger the scale of the fields is, the higher the water intake cost. (2) Weeding must be implemented frequently, and land bearing capacity reduces. (3) Other issues are related to the collapsing ridge of a paddy field and the burden of fertilizer spraying, and these factors result in a decrease in work efficiency. (4) Finally, flooding paddy fields compete with other land uses, such as wheat cultivation (Mineta et al. 2004).

If it is difficult to flood an entire paddy field, then creating a fish-retreat ditch at the beginning of establishing a wetland should be considered. On the other hand, issues related to creating a fish-retreat ditch involve the following: (1) The paddy field area decreases, and the yield decreases. (2) Installation of a paddy fishway with a fish-retreat ditch is required, and the installation cost is high. (3) Labor and technical skills are required to create a fish-retreat ditch and paddy fishway. (4) Maintenance, such as weeding in the fish-retreat ditch, is required.

Currently, the area of paddy fields is decreasing due to the aging and depopulation of rural areas, and the conversion of paddy fields is progressing due to paddy farmland

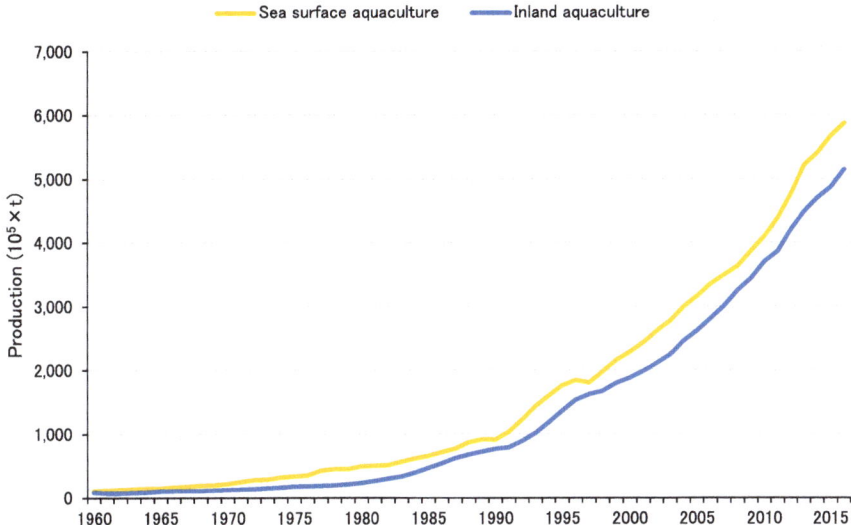

Fig. 12.14 Changes in world aquaculture production. *Source* Illustrated from FAO, FishStat (Capture Production, Aquaculture Production), http://www.fao.org/fishery/statistics/en, Accessed November 3, 2021

consolidation. For aquatic animals such as fish to survive in the future under such circumstances, it is necessary for farmers to be willing to create permanent water areas such as fish-retreat ditches. To secure a permanent water area, Sado City has ensured the habitat of aquatic animals by introducing a certification system for rice cultivation. Since this area is the last breeding ground for wild *Ciconia boyciana* in Japan, it is desirable to introduce a rice cultivation certification system (for example, *Ciconia boyciana* township rice) in Obama City to create a rich habitat for the ecosystem.

Box 12.1 Fish Farming in Paddy Fields

Loach, *Misgurnus anguillicaudatus*, and catfish, *Silurus asotus*, are drainage-paddy field fishes, and they ascend from the drainage canal to the paddy field to spawn in the paddy field. Consolidating paddy fields that have become abandoned cultivated land in one place and cultivating catfish should be conducted. There are many catfish in the Asian monsoon zone. Catfish are very large and enjoyable to eat. In Japan, they are attracting attention as an alternative ingredient to eel, *Anguilla japonica*.

Global sea surface aquaculture and inland aquaculture productions are growing exponentially (Appendix Fig. 12.14). Considering aquaculture production by country, the increase in China and Indonesia is remarkable, with China producing 63.72 million tons, or 58% of the world's production,

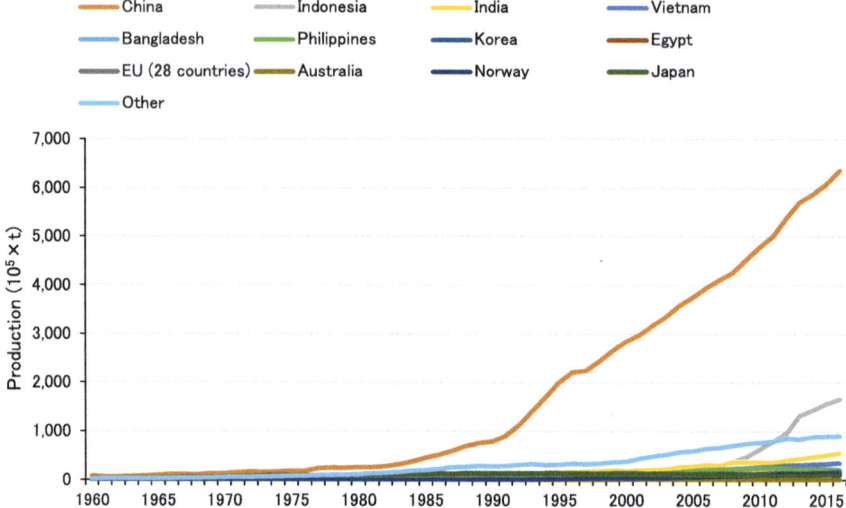

Fig. 12.15 Changes in world aquaculture production by country. *Source* Illustrated from FAO, FishStat (Capture Production, Aquaculture Production), http://www.fao.org/fishery/statistics/en, Accessed November 3, 2021

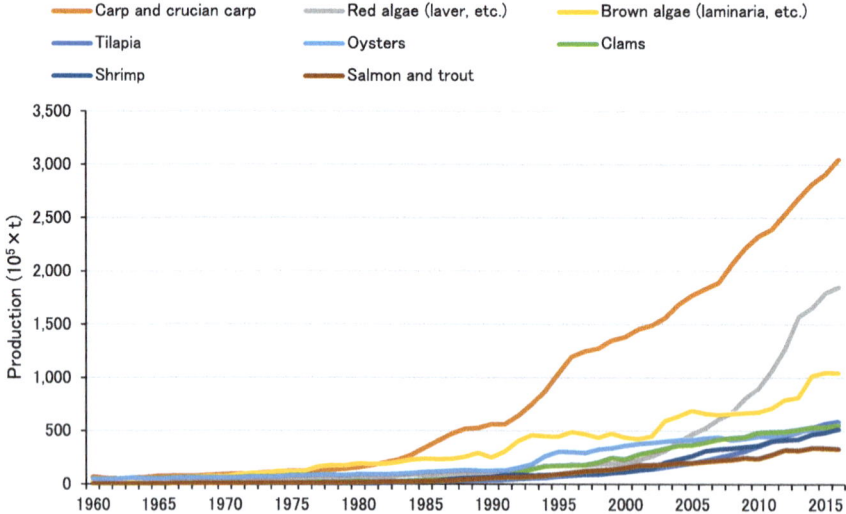

Fig. 12.16 Changes in world aquaculture production by fish. *Source* Illustrated from FAO, FishStat (Capture Production, Aquaculture Production), http://www.fao.org/fishery/statistics/en, Accessed November 3, 2021

and Indonesia producing 16.62 million tons, or 15% of the world's production (Appendix Fig. 12.15). In terms of fish, carp and crucian carp accounted for the largest produced amount at 30.54 million tons, accounting for 28% of the global total, followed by red algae at 18.47 million tons and brown algae at 10.51 million tons. The increase in fish species production is remarkable (Appendix Fig. 12.16).

Therefore, catfish are considered to be one of the important aquaculture fish in the world. There is no custom of eating the catfish in Japan, but there is a custom of eating eel. Considering the current situation that the biomass of eels will decrease in the future, it is desirable to cultivate and eat catfish (Appendix Fig. 12.17).

Appendix

See Figs. 12.14, 12.15, 12.16 and 12.17.

Fig. 12.17 Catfish, *Silurus asotus*, in Japan (*Photo* by Akira Matsui). Notes: Photo date: June 26, 2021. The author breeds these catfish species

References

Dotsu Y (1955) The life of a Goby, *Chaenogobius urotaenia* (Hilgendorf) (in Japanese with English Abstract). Sci Bull Faculty Agric 15: 367–374. https://doi.org/10.15017/21383

Echizen City (2020) Material explanation of terms (in Japanese). https://www.city.echizen.lg.jp/off ice/060/020/syokutonou_p_d/fil/shiryouhen.pdf. Accessed 26 June 2021

Hata K (1999) Field experiment on the migration of fishes to an idle rice paddy with a small fishway (in Japanese). J Jpn Soc Irrig Drainage Reclam Eng 67: 497–502. https://doi.org/10.11408/jjsidr e1965.67.5_497

Hiyori Y, Fujinaga Y, Mizutani M, Tawa K, Sagawa S (2016) The effect of permanent water-filled ditches installed in paddy fields for a foraging habitat of oriental white storks; a case of the ecosystem preservation of the paddy field in Echizen City, Fukui Prefecture (in Japanese with English Abstract). Reintroduction 4:29–36

Hiyori Y, Fujinaga Y, Mizutani M, Tawa K, Sagawa S (2017) The aquatic fauna in water-filled ditches developed in paddy field during the early summer and summer seasons (in Japanese with English Abstract). Reintroduction 5:39–46

Ishima T, Murakami H, Takahashi Y, Iwamoto Y, Takanose Y, Sekijima T (2016) Method for creation of swale as a tactic for improving habitat of freshwater fish in conventional paddy field (in Japanese with English Abstract). Ecol Civ Eng 19:21–35. https://doi.org/10.3825/ece.19.21

Ishino K (1989) *Gymnogobius urotaenia* (in Japanese). In: Kawanabe H, Mizuno N, Hosoya K (eds) Freshwater fishes of Japan. Yama-kei Publishers Co., Ltd, Tokyo, pp 618–620

Iwabuchi S (2003) Significance and utilization of flooded paddy fields in winter—recycling technology that supports productivity while maintaining the ecosystem–(in Japanese). Rural Environ 19:50–59

Kubota Z (1961) Ecology of the Japanese loach, *Misgurnus anguillicaudatus* (CANTOR)—IV. Growth and fatness (in Japanese with English Abstract). J Shimonoseki Coll Fish 11: 213–234

Kurechi M (2007) Restoring rice paddy wetland environments and the local sustainable society— project for achieving co-existence of rice paddy agriculture with waterbirds at Kabukuri-numa, Miyagi Prefecture, Japan. Glob Environ Res 11:141–152

Kurechi M (2016) Effect of winter paddy field on the conservation of birds (in Japanese). Agric Hortic 91:87–97

Matsui A, Satoh M (2004) Distribution of aquatic animals in the drainage systems created by paddy farmland consolidation in Shimodate City, Ibaraki Prefecture, Japan (in Japanese with English Abstract). Jpn J Conserv Ecol 9:153–163. https://doi.org/10.18960/hozen.9.2_153

Matsui A (2009) Growth of several fish and dragonfly species in the drainage system of a consolidated paddy field (in Japanese with English Abstract). Jpn J Conserv Ecol 14:3–11. https://doi.org/10.18960/hozen.14.1_3

Matsui A (2021) Fish community in a fish-retreat ditch created in a paddy irrigation and drainage system near the sea (in Japanese with English Abstract). Jpn J Conserv Ecol 26:165–175. https://doi.org/10.18960/hozen.2015

Matsui A (2022) An overwintering ground for fish in a fish-retreat ditch created in paddy irrigation and drainage system near the sea (in Japanese with English Abstract). Ecol Civ Eng 24:245–258. https://doi.org/10.3825/ece.21-00018

Mineta T, Kurita H, Ishida K (2004) Potential of winter-flooded rice field in regard to farming and the multifunctionality—analysis of status by questionnaire and interview survey to operative farmers—(in Japanese with English Abstract). Trans Rural Plan 6:61–66. https://doi.org/10.2750/arp.23.23-suppl_61

Miyaji D, Kawanabe H, Mizuno N (1963) Colored illustrations of the freshwater fishes of Japan (in Japanese). Hoikusha Co., Ltd, Osaka

Miyasato K (2009) "Island that coexists with *Nipponia nippon*" Sado's agricultural and rural development-From "creating a feeding ground" to "paddy fields that live with *Nipponia nippon*"— (in Japanese). http://seneca21st.eco.coocan.jp/working/miyasato/19_04_1.html. Accessed 26 June 2021

Nakajima J, Uchiyama R (2017) Loaches of Japan—natural history and culture—(in Japanese). Yama-kei Publishers Co., Ltd, Tokyo

Nakamura M (2002) About the growth and maturity of *Acanthogobius flavimanus* (in Japanese). Rep Freshwater Fish Exp Station, Ibaraki Prefecture 37:29–34

Nishida K, Fujii C, Minagawa A, Senga Y (2006) Research on migration and dispersal range of freshwater fish that reproduce in temporary water area—Case study of Mukojima-channel in Hino-city and Fuchu-channel in Kunitachi-city, Tokyo—(in Japanese with English Abstract). Trans Jpn Soc Irrig Drain Reclam Eng 244:151–163. https://doi.org/10.11408/jsidre1965.200 6.553

Nishikawa U (2015) Environmentally friendly farming in Japan: introduction. In: Nishikawa U, Miyashita T (eds) Social-ecological restoration in paddy-dominated landscapes, Ecological research monographs. Springer, Tokyo, pp 69–86. https://doi.org/10.1007/978-4-431-55330-4_5

Sado City (2021) Globally important agricultural heritage systems "Sado's Satoyama in coexistence with *Nipponia nippon*" Agriculture is a culture (in Japanese). https://www.city.sado.niigata.jp/site/giahs/. Accessed 26 June 2021

Sakai H (1989) *Tribolodon hakonensis* (in Japanese). In: Kawanabe H, Mizuno N, Hosoya K (eds) Freshwater fishes of Japan. Yama-kei Publishers Co., Ltd, Tokyo, pp 259–264

Sakai K, Katayama S, Iwata T (2000) Life history of the Japanese common goby, *Acanthogobius flavimanus* in the Matsushima Bay (in Japanese). Bull Miyagi Prefectural Fish Res Dev Center 16:85–92

Sato T, Sato M, Inagaki M, Sato T, Anjitsu C, Tsuchida K, Misawa S (2008) Ascending migration of loach *Misgurnus anguillicaudatus* affected by water management practices of paddy fields and installation conditions of corrugated pipes as water fishways to paddy fields (in Japanese). J Rural Plan Assoc 26:434–441. https://doi.org/10.2750/arp.26.434

Shiga Prefecture (2018) Fish cradle paddy project (in Japanese). https://www.pref.shiga.lg.jp/ippan/shigotosangyou/nougyou/nousonshinkou/18537.html. Accessed 26 June 2021

Suzuki M, Mizutani M, Goto A (2004) Effects of connection of paddy fields, ditch and stream through small-scale fishways on fish fauna (in Japanese with English Abstract). Trans Jpn Soc Irrig, Drainage Reclam Eng 234:641–651. https://doi.org/10.11408/jsidre1965.2004.641

Tsuji K (1989) *Acanthogobius flavimanus* (in Japanese). In: Kawanabe H, Mizuno N, Hosoya K (eds) Freshwater fishes of Japan, 624 Yama-kei Publishers Co., Ltd., Tokyo

Washitani I (2007) Restoration of biologically-diverse floodplain wetlands including paddy fields. Glob Environ Res 11:135–140

Part V
Wetland Development

Chapter 13
Improvements to Enrich the Paddy Irrigation and Drainage Canal Systems

Abstract The purpose of this study was to propose selective and step-by-step conservation measures for the irrigation system in consideration of biological conservation. To carry out this study, fish and dragonfly habitats were surveyed and analyzed regularly from April 2001 to June 2003 in the irrigation and drainage canal systems of a consolidated paddy field, Shimodate City (now Chikusei City), Ibaraki Prefecture, Japan. The proposed conservation measures are followed from the simplest one to the most complex one: (1) install a discharger for fish movement from the irrigation canal system to the drainage canal system, (2) install a fishway between the main canal and the farm drain, (3) adopt a shallow farm drain by separating underground wastewater and (4) pass water through the farm drain during the nonirrigation season. In particular, if winter water enriches the paddy field ecosystem, then consequently, the environment of the entire region including the river ecosystem connected to the paddy field ecosystem will also be enriched.

Keywords Biological conservation · Dragonfly · Fish · Irrigation and drainage canal systems · Paddy farmland consolidation · Winter water

13.1 Introduction

On the basis of the survey results revealed by Chaps. 10, 11 and 12, the irrigation and drainage canal systems are considered to conserve biodiversity. Because the measures implemented to conserve aquatic animals can vary depending on the organism, a variety of selective and gradual approaches are proposed to preserve the fish and dragonflies in a consolidated paddy field. Four conservation measures are proposed below (Table 13.1).

This chapter is a revised version of Matsui and Satoh (2005). Copyright 2005 The Japanese Society of Irrigation, Drainage and Rural Engineering, https://doi.org/10.11408/jjsidre1965.73.4_277, Accessed November 28, 2021.

© The Author(s), under exclusive license to Springer Nature Singapore Pte Ltd. 2022
A. Matsui, *Wetland Development in Paddy Fields and Disaster Management*,
https://doi.org/10.1007/978-981-19-3735-4_13

239

Table 13.1 Selective and gradual conservation measures and expected effects for aquatic animals

Conservation measure	Discharge paths from irrigation canal system to drainage canal system	Fishways between main canal and farm drain	Shallow farm drain with an underdrain system	Passing water through a shallow farm drain for a nonirrigation season	Expected effect
Current state	×	×	×	×	-
A	○	×	×	×	Movement of fish from irrigation canal system to drainage canal system
B	○	○	×	×	Conservation of irrigation–drainage fish
C	○	○	○	×	Conservation of drainage–paddy field fish
D	○	○	○	○	Conservation of dragonflies

Source Reprinted from Matsui and Satoh (2005). Copyright 2005 The Japanese Society of Irrigation, Drainage and Rural Engineering

13.2   Selective and Gradual Conservation Measures for Aquatic Animals

13.2.1   Conservation Measure A

Discharge paths from the irrigation canal system to the drainage canal system are installed for fish refuge. Based on the current status of consolidated paddy fields, irrigation and drainage canals are generally divided. Fish that have flowed into an irrigation canal from a river will die in the canal when irrigation water is stopped (Fig. 13.1). As a countermeasure, the fish that exist in the irrigation canal need to be removed and placed at the intersection of the irrigation and drainage canal systems before the water is stopped (Fig. 13.2). In this case, the water should be flowing in the drainage canal system in the nonirrigation season.

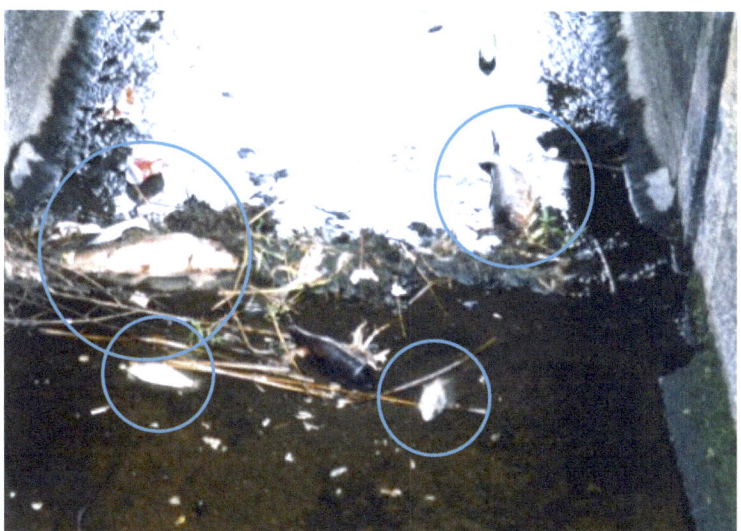

Fig. 13.1 Fish that died in an irrigation canal system (*Photo* by Akira Matsui)

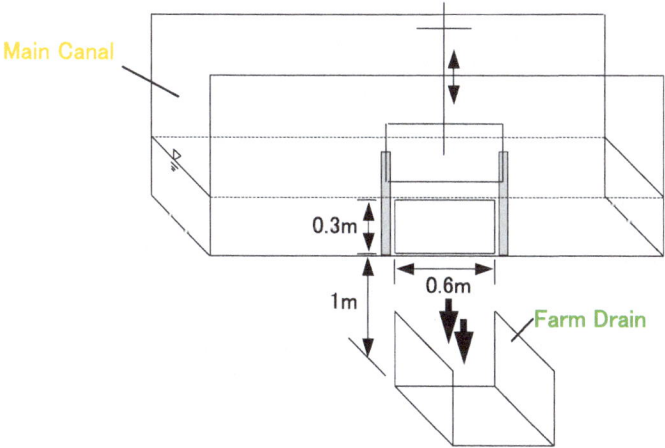

Fig. 13.2 Discharge paths from an irrigation canal system to a drainage canal system

13.2.2 *Conservation Measure B*

Fishways between a main canal and a farm drain are established, and irrigation water is discharged (Fig. 13.3). Thus, it will be possible for the irrigation–drainage fish to move between the irrigation and drainage canal systems in the irrigation season. If the flow rate at the farm drain increases by discharging water, then the fish in the main drain are likely to be attracted to the farm drain.

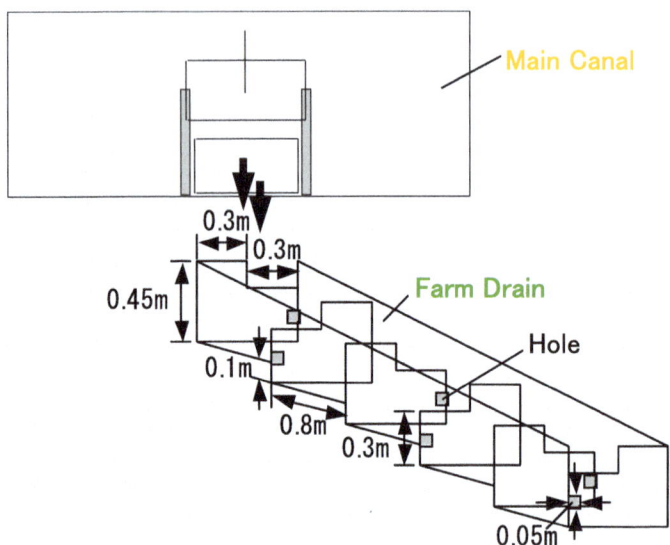

Fig. 13.3 Fishways between a main canal and a farm drain. *Source* Modified from Hata (2005). Copyright 2005 Rural Culture Association Japan

13.2.3   Conservation Measure C

To separate surface drainage and subsurface drainage, shallow farm drains are constructed (Fig. 11.15). This proposed measure ensures that the drainage–paddy field fish, which live mainly in the drainage canal system, can move directly and easily into the paddy fields in the irrigation season for spawning. By introducing shallow farm drains, the spawning grounds for the drainage–paddy field fish dramatically increase. For this measure, the fishway from conservation measure B is moved to the intersection at the farm drain and the main drain.

Regarding the adoption of the shallow farm drain, the farmland consolidation that Shinzawa and Koide (1963) proposed to prevent a reduction in irrigation area and increase labor savings of slope management needs to be reevaluated from the modern biological conservation perspective. The proposal in this study adopts the shallow farm drain by installing an underdrain system for subsurface water drainage, resulting in a variety of physical and economic benefits. Even if the shallow farm drain is constructed as an earthen canal, the area of the earthen canal is so small that the administrative burden will be limited. Furthermore, as it is not necessary to install a fishway on each individual paddy field, it is possible to reduce the number of fishways significantly.

13.2.4   Conservation Measure D

The shallow farm drain should have flowing water in the nonirrigation season. Fish can move to wintering places where the water depth is sufficient, while most dragonflies, such as *Orthetrum albistyrum speciosum* and *Calopteryx atlata,* will overwinter in the form of larvae at certain places. Hence, the mobility of dragonflies is so minimal that flowing water in the nonirrigation season protects them from dryness.

Among these conservation measures, conservation measure D is affected by the water rights issue. Therefore, implementing this measure is likely to take time as the water rights issue must be resolved. Conservation measures B and C, which would be constructed together, are efficient and expensive. Conservation measure A is simple and effective and can be conducted immediately. However, conservation measure A is only the first step, and conservation measures B, C and D are required in the future.

13.3   Water Flow Method in the Nonirrigation Season

Flowing water in the nonirrigation season is required in the farm drain as well as in the lateral and main drains. However, this action must be carefully implemented so that biodiversity is not impaired by the process of ensuring water flow at all farm drains. In terms of the water flow rate at the farm drain in the nonirrigation season in this study, the average water depth was approximately 5 cm and almost stopped at FD1 (Fig. 10.9a), where *Orthetrum albistyrum speciosum* was caught in large numbers (Table 10.4). Thus, a high water flow rate will not be needed.

To ensure running water in the nonirrigation season, the water taken from the river must be sent through the irrigation canal system to the drainage canal system. Because water intake is based on securing the maintenance river flow in winter, it is difficult to secure new water rights at present. However, water will be used and reduced in the river with no consumption (Satoh 2002). It is inevitable that the river flow rate in the section from the intake point to the reduced water inflow point is reduced. When paddy ecosystems are biologically rich, the river is also considered to be healthy ecosystems because of the connection between the river and irrigation and drainage canals. Therefore, unless water reduction has a particularly large impact on river ecology, it would be preferable for the entire area, including the river environment, to ensure small amounts of water in the irrigation and drainage canals in the nonirrigation season.

Even in the irrigation season, it is necessary to discharge into the drainage canal system from the irrigation canal system to connect the main canal and the farm drain with flowing water; however, this is an invalid discharge of irrigation water, even though the discharge will flow down into the drainage canal system without being consumed. If the discharge is reused as irrigation water in the appropriate place, then there will be no water consumption problem. Given this is the area of reuse, it is desirable that the irrigation and drainage canals are close in the downstream portion.

13.4   Water Flow Plan for the Irrigation and Drainage Canal Systems

Based on the investigation of the flying distance of adult dragonflies from a particular breeding ground, *Cercion sieboldii* and *Ischnura asiatica* fly 1.2–1.3 km, and *Crocothemis servilia mariannae* fly 1.0–1.1 km (Moriyama et al. 1990). When providing dragonfly breeding grounds within their range, it is important to ensure gene immobilization does not occur so that gene exchange can occur.

For dragonfly conservation, it is sufficient to ensure year-round water flow within a maximum interval of approximately 1.0 km in the shallow farm drain, which is proposed in Fig. 11.15. If aquatic insects, e.g., dragonflies, are numerous at shallow farm drains, then they are also providing effective fish conservation because they are providing food resources for the fish.

Additionally, when draining in the nonirrigation season, determining whether the water flowing in the drains should move directly from the main canal to the farm drain or through the paddy field should be based on the conditions of each district. In this study area, herons were observed throughout the year and foraged in the drainage canal in the nonirrigation season. Hence, if the paddy fields are submerged, then they can serve as possible foraging locations for herons. On the other hand, it is a concern that the earth yield strength would be reduced in paddy fields by flooding them in the winter. In the future, in consideration of these advantages and disadvantages, it will be necessary to determine whether to flood the paddy fields with flowing water in the nonirrigation season.

13.5   Paddy Fields as Social Assets

In this study, based on the analysis of the ecological distribution of aquatic animals, conservation measures were selected according to the target organisms and the ability of the measures to be implemented in a step-by-step manner, and the measures are shown in Table 13.1. These proposed measures have the following features: They can be implemented to take corrective action depending on the target organisms and the available financial resources for conservation in a consolidated paddy field. Given

Table 13.2 Type of paddy fishways

No	Fishway type			Photo
A	Directly connected to paddy field	Partition type	Chidori-X	
B			Half-cone	
C		Rough surface type	Wave	
D	Two-layer drainage system			

(continued)

the need for increased environmental conservation in the future, whether Japan's paddy fields are sustained and accepted as social assets depends on the implementation of effective and efficient biological conservation measures that also result in high productivity levels. These proposals are based on the analysis of the ecological distribution of aquatic animals and are expected to advance the discussion of future conservation measures.

Table 13.2 (continued)

No	Fishway type	Photo
E	Waterway weir	

Source Modified from the Rural Environment Division, Rural Policy Department, Rural Promotion Bureau, Ministry of Agriculture, Forestry and Fisheries of Japan, https://www.maff.go.jp/j/nousin/kankyo/kankyo_hozen/pdf/gyodou.pdf, Accessed November 3, 2021 (in Japanese)

Box 13.1 Type of Paddy Fishways

The role of a paddy fishway is to conserve the fish that use paddy fields for spawning, growing and overwintering. In recent years, the number of fish near paddy fields has been decreasing. Because there are height differences between drainage canals and paddy fields in consolidated paddy fields, fish cannot ascend to the paddy fields. Therefore, efforts are being implemented nationwide to connect paddy fields and drainage canals with paddy fishways so that fish can ascend to paddy fields from drainage canals. From various survey results, it has been shown that the installation of a paddy fishway effectively enables fish such as loach, catfish, carp and crucian carp to ascend to a paddy field and descend from a paddy field. Appendix Table 13.2 shows the types of the paddy fishways.

A The features of the Chidori-X fishway are as follows: The upper part of the partition wall is tilted in the transverse direction, and the partition walls are arranged in a staggered pattern. Since the pool is created by the partition wall, the overflow depth can be secured even with a small flow rate. Because it has a variety of flows, it can accommodate many fish species. It must be installed in each section. Sediment accumulates in the pool.

B The features of the half-cone fishway are as follows: The structure is the same as that of the Chidori-X fishway, but the partition wall of the overflow part is cylindrical to smooth the overflow. There is minimal separation of flow, and a pool can be formed. Various flows can occur even with a low flow rate. Therefore, this type of fishway can be used for many fish species. There is little sedimentation. It must be installed in each section, and the installation cost is high.

C The features of the wave fishway are as follows: The bottom surface is uneven to reduce the flow so that fish can move easily. The flow rate and water depth in the fishway may be minimal. This fishway is inexpensive, lightweight and easy to manage. It must be installed in each section. The target fish species are limited to fish species with a low withers height.

D The features of the two-layer drainage system fishway are as follows: The upper soil canal is used for fish habitat and rainwater drainage, and the lower culvert is used for culvert drainage and heavy rain drainage. In the upper soil canal, there is no height difference between the drainage canal and the paddy field. All fish species that live in the drainage canal can survive in the upper soil canal. There is no need to install it in each section. Installation cost is high. The two-layer drainage system fishway is applied on low and gentle slopes. Since a soil canal is constructed, it is necessary to establish a maintenance system in advance.

E The features of the waterway weir fishway are as follows: The waterway weir fishway eliminates the difference in water level between a paddy field and a waterway during irrigation season or fish spawning. Temporary management efforts are sufficient. All fish species that live in waterways can survive these fishways. There is a risk of slope erosion, and it is necessary to remove the weir board in the event of a flood.

Appendix

See Table 13.2.

References

Hata K (2005) How medaka survives the crisis—making fish ascend to the paddy field—(in Japanese). Rural Culture Association Japan, Tokyo

Matsui A, Satoh M (2005) A proposal of selective conservation of aquatic animals in irrigation and drainage canals of a consolidated paddy field (in Japanese). J Japanese Soc Irrig, Drainage Reclam Eng 73: 277–280. https://doi.org/10.11408/jjsidre1965.73.4_277

Moriyama H, Iijima H, Harada N (1990) A favorable distribution pattern of aquatic biotopes in terms of the dispersal ability of dragonflies (in Japanese). Man Environ 15:2–15

Satoh M (2002) Vision; to manage the region's water (in Japanese). J Japanese Soc Irrig, Drainage Reclam Eng 70:797–798. https://doi.org/10.11408/jjsidre1965.70.9_797

Shinzawa K, Koide S (1963) Land readjustment of arable land (in Japanese). Iwanamishoten Publishers Co., Ltd, Tokyo

Chapter 14
Multiple Effects of Wetland Development

Abstract Abandoned cultivated lands should be consolidated in one place, and these lands should be developed into wetlands. Since wetlands are permanent waters, fish and dragonflies can live without the danger of dry conditions. Wetlands also help reduce the flow of the main river in a flood disaster. Thus, wetlands have multiple functions. These functions include biodiversity conservation and flood control. In addition, children can enhance their environmental knowledge when visiting wetlands, and this wetland function is the environmental education function. In addition, edible plants such as lotus root can be grown in wetlands, fulfilling food supply and landscape conservation functions. Fishing, walking and jogging in wetlands are recreational and sport-related functions.

Keywords Biodiversity conservation · Environmental education · Flood control · Food supply · Landscape conservation · Multiple functions · Recreation · Sports · Wetland

14.1 Introduction

According to the author's research, there are many aquatic animals, such as fish and dragonflies, in paddy field systems, especially in the drainage systems. In addition, the irrigation system plays a role as a habitat. However, the biodiversity conservation function of the irrigation system is inferior to that of the drainage system because it has a pipeline or three concrete surfaces. In this study, the fish-retreat ditch was used as a habitat, a spawning ground and an overwintering ground for fish that mainly lived in the drainage system.

On the other hand, paddy fields are not being cultivated due to the aging of farmers and the depopulation of rural areas. Abandoned cultivated lands are increasing, and maintenance of paddy irrigation and drainage systems is not being carried out. As a result, the biodiversity conservation function of paddy irrigation and drainage systems is not occurring. In addition, wetlands that used to be approximately 2110 km^2 in the Meiji and Taisho eras now cover 820 km^2, and this value represents an approximate 60% loss of wetlands (Geospatial Information Authority of Japan 2000).

© The Author(s), under exclusive license to Springer Nature Singapore Pte Ltd. 2022
A. Matsui, *Wetland Development in Paddy Fields and Disaster Management*,
https://doi.org/10.1007/978-981-19-3735-4_14
249

Therefore, consolidating abandoned cultivated land into one place and converting it to wetlands are proposed. Since wetlands are permanent waters, fish and dragonflies can live without the danger of dry conditions. Wetlands also help reduce the flow of main rivers during water disasters.

If there are open levees, then wetlands are better positioned in the immediate vicinity of these open levees. If fish-retreat ditches are placed at regular intervals in addition to wetlands, then a network of 'wetlands-fish-retreat ditch-paddy field' will be created, and it is expected that the biodiversity conservation function of the paddy irrigation and drainage system will be enriched more than ever. Furthermore, if this approach leads to disaster prevention and mitigation, this approach is addressing three issues at once.

14.2   Flood Control

By creating wetlands, it is possible to store water during floods. As a result, the damage caused by heavy rain is reduced. The conversion of wetlands into flood control basins is part of watershed control; thus, wetlands are green infrastructure (Fig. 14.1).

Table 14.1 summarizes the advantages and disadvantages of gray infrastructure (man-made structures) and green infrastructure. Infrastructure with man-made structures can achieve a single function with high accuracy that serves a specific and clear purpose. The greatest advantage of this type of infrastructure is that it can accurately provide the performance required by society. In addition, various economic effects, such as short-term job creation in the region where it is constructed, may be one of the

Fig. 14.1 Components of green infrastructure. *Source* Modified from the Ministry of Land, Infrastructure, Transport and Tourism of Japan, https://www.mlit.go.jp/common/001179745.pdf, Accessed November 3, 2021 (in Japanese)

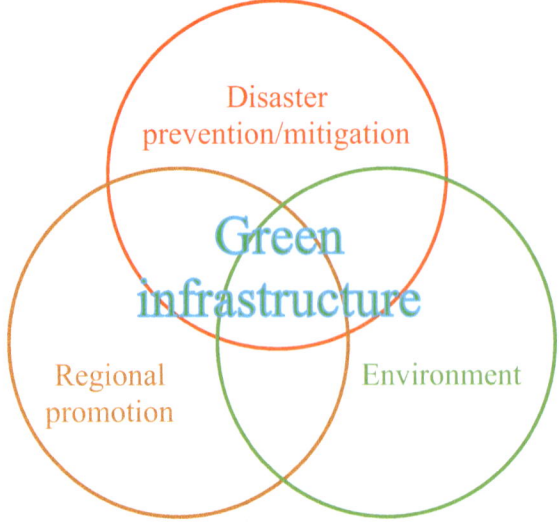

Table 14.1   Advantages and disadvantages of gray infrastructure and green infrastructure

	Gray infrastructure	Green infrastructure
Demonstration of a single function	◎	△
Multifunction	△	◎
Adaptive management with uncertainty	×	○
Avoiding environmental burden	×	◎
Short-term job creation/economic effects on the region	◎	△
Long-term job creation/economic effects on the region	△	○

Source Modified from the Science Council of Japan, http://www.scj.go.jp/ja/info/kohyo/pdf/kohyo-22-t199-2.pdf, Accessed November 3, 2021 (in Japanese)

reasons why a region welcomes gray infrastructure. On the other hand, the greatest advantage of green infrastructure is that it maintains and creates multiple-use spaces that can provide diverse ecosystem services that also contribute to the conservation of biodiversity. Adaptive management also has the advantage of addressing uncertainty.

14.3   Biodiversity Conservation

Wetlands serve as overwintering grounds for fish and dragonflies and provide biological conservation effects. Fish inhabit, spawn and overwinter in wetlands. As a result, this increase in fish allows birds to prey on them (Fig. 14.2).

14.4   Environmental Education

Children can enhance their environmental education in wetlands during elementary school environmental studies (Fig. 14.3). Elementary school students learn that wetlands are flooded year round and that fish and other animals live there. They are effective as wintering grounds for fish, especially in winter when paddy fields are drained of water.

14.5   Food Supply and Landscape Conservation

It is possible to grow edible plants such as lotus roots in wetlands. In addition, these plants are expected to absorb nutrients and purify eutrophic water, while also providing a good landscape (Fig. 14.4).

Fig. 14.2 Food chain. *Source* Reprinted from the Ministry of Agriculture, Forestry and Fisheries of Japan, https://www.maff.go.jp/j/nousin/noukan/nougyo_kinou/pdf/adult_zentai.pdf, Accessed November 3, 2021 (in Japanese)

14.6   Recreation and Sports

One can enjoy fishing in a wetland. If one cover an area around a wetland with soil, then walking and jogging can occur there (Fig. 14.5). In Japan and the world, society is aging, and there is a declining birth rate. Elderly people should exercise by walking to maintain their health. Jogging is also the best exercise for young people. It is very safe to jog around a wetland because there are no cars.

Box 14.1 Multiple Functions of Agricultural and Rural Areas

Agricultural and rural areas provide multiple functions in addition to the food and other products resulting from agricultural production activities in rural areas, and these additional functions include national land conservation, water source recharge, natural environment conservation, good landscape formation and cultural transmission. Thus, these are multifaceted functions that are provided in addition to the function of supplying agricultural products (Appendix Fig. 14.6).

However, agriculture is no longer in operation because the agricultural and rural areas of Japan have declined due to the effects of depopulation and aging. The agricultural and rural areas with these multiple functions must be conserved in the future.

Fig. 14.3 Environmental education. *Notes* Photo date: June 30, 2021. The author is explaining wetlands

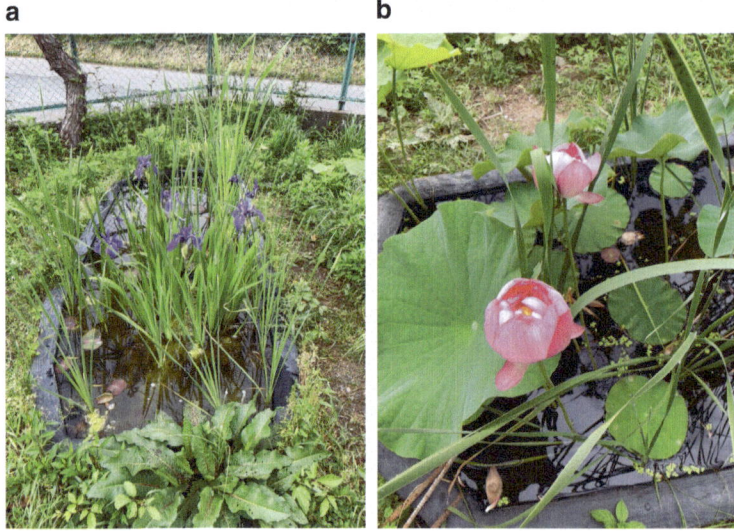

Fig. 14.4 Landscape conservation: **a** *Iris laevigata* and **b** *Nelumbo nucifera* (*Photo* by Akira Matsui). *Notes* **a** Photo date: May 17, 2022. **b** Photo date: July 14, 2021. This wetland is in the author's garden

Fig. 14.5 Walking or jogging course around a wetland: **a** ground plan and **b** sectional plan

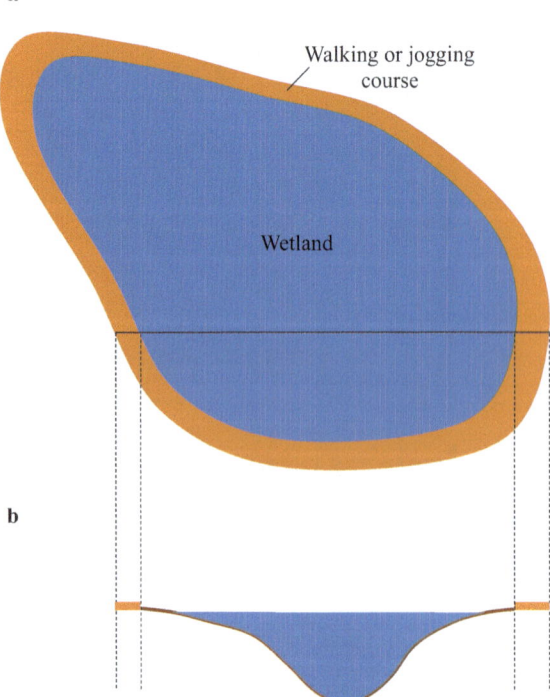

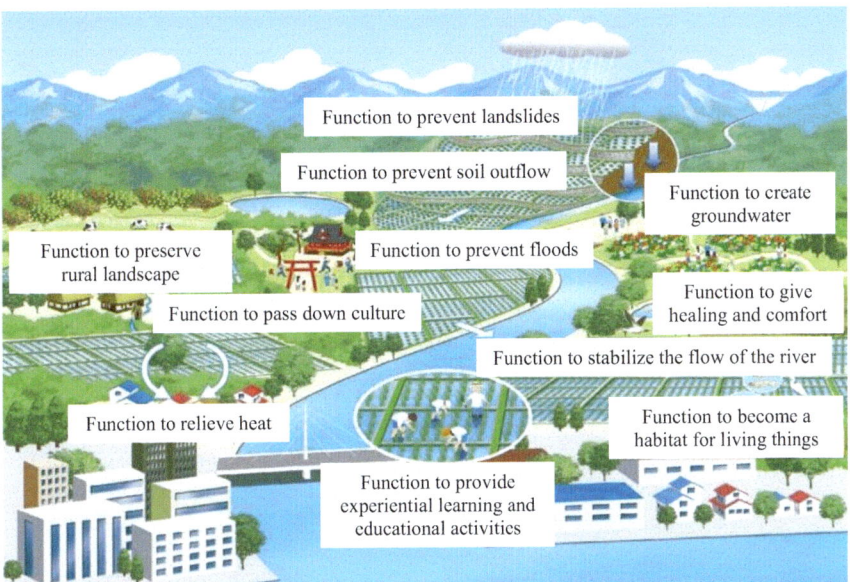

Fig. 14.6 Multiple functions of agricultural and rural areas. *Source* Modified from the Ministry of Agriculture, Forestry and Fisheries of Japan, https://www.maff.go.jp/j/nousin/noukan/nougyo_kinou/img/zentai02.jpg, Accessed November 3, 2021 (in Japanese)

Appendix

See Fig. 14.6.

Reference

Geospatial Information Authority of Japan (2000) Survey results of changes in wetland area throughout Japan. https://www.gsi.go.jp/kankyochiri/shicchimenseki2.html. Accessed 19 Nov 2021 (in Japanese)

Chapter 15
Wetland Management

Abstract Questions such as what is the desired size of wetland and should a very large wetland or many small wetlands be created were considered when determining the wetland size needed to create an environment for higher consumer, e.g., Oriental white stork, *Ciconia boyciana*. There needs to be large wetlands with 1000 m diameters and small wetlands with 100 m diameters. The larger wetlands should be located at 3000 m intervals and the smaller wetland should be located at 500 m intervals. To create the wetlands, it would be beneficial to combine abandoned cultivated land. In the event of large floods, water can be stored in the wetlands, as well as in an open levee, drainage canal systems, fish-retreat ditches and paddy fields. It is possible that the damage caused by large water disasters in recent years has expanded due to the loss of autonomy of the residents in the basin. For the hydraulic control in this basin to be successful, the autonomy of previously existing basin residents must be restored. The participants in this effort are basin residents, municipal governments and academia.

Keywords Academia · Autonomy · Basin resident · Interval · Municipal government · Size · Wetland

15.1 Sizes and Sites of Wetlands

Questions such as what is the desired size of wetland and should a very large wetland or many small wetlands be created were considered when determining the wetland size needed to create an environment for higher consumer, e.g., Oriental white stork, *Ciconia boyciana*. Since the home range of *C. boyciana* will need a large circle approximately 3000 m in diameter, the wetland needs to be a large wetland circle approximately 1000 m in diameter and a small wetland circle approximately 100 m in diameter (Fig. 15.1a). Larger wetlands should be located at approximately 3000 m intervals, and smaller wetlands should be located at approximately 500 m intervals (Fig. 15.1b).

Fish-retreat ditches correspond to small wetlands. The best permanent waters are rivers. By linking wetlands, fish-retreat ditches and rivers, it is possible to create an environment where *Ciconia boyciana* can live and spawn. Nishida et al. (2006)

© The Author(s), under exclusive license to Springer Nature Singapore Pte Ltd. 2022
A. Matsui, *Wetland Development in Paddy Fields and Disaster Management*,
https://doi.org/10.1007/978-981-19-3735-4_15

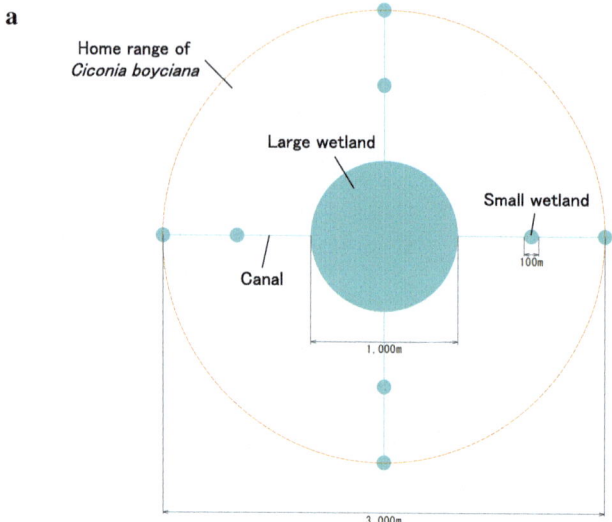

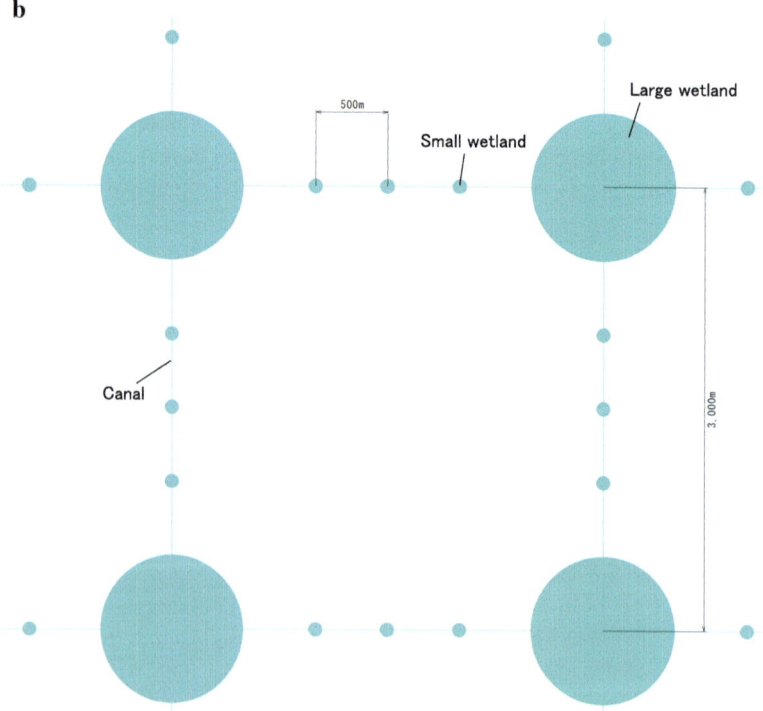

Fig. 15.1 **a** Size and **b** site of wetland

reported that the maximum movement distance of *Misgurnus anguillicaudatus* that use temporary waters for breeding is 500 m. Moriyama et al. (1990) reported that the flying distance of adult dragonflies from a particular breeding ground, *Cercion sieboldii* and *Ischnura asiatica* fly 1.2–1.3 km, and *Crocothemis servilia mariannae* moves 1.0–1.1 km. Therefore, an arrangement interval of 500 m for the wetlands is proposed.

15.2 Importance of Regional Ecosystems

Global warming is steadily progressing, and large floods are occurring frequently. On the other hand, paddy ecosystems have multiple functions in Japan, which is located in the Asian monsoon region. The multiple functions of paddy fields are flood control, biodiversity conservation, environmental education, food supply, landscape conservation, recreation and sports. Abandoned cultivated land is increasing in Japan; therefore, it would be beneficial to combine abandoned cultivated land to create wetlands.

Regional ecosystems consist of paddy ecosystems and river ecosystems (Fig. 15.2). Paddy ecosystems consist of paddy fields, irrigation and drainage canal systems, fish-retreat ditches and wetlands. In the event of large floods, water can be stored in wetlands, as well as an open levees, drainage canal systems, fish-retreat ditches and paddy fields. River ecosystems are permanent waterbodies and rich in aquatic animals. Therefore, it is desirable that river ecosystems and paddy ecosystems are aligned. In the future, we must maintain the paddy ecosystems in the Asian

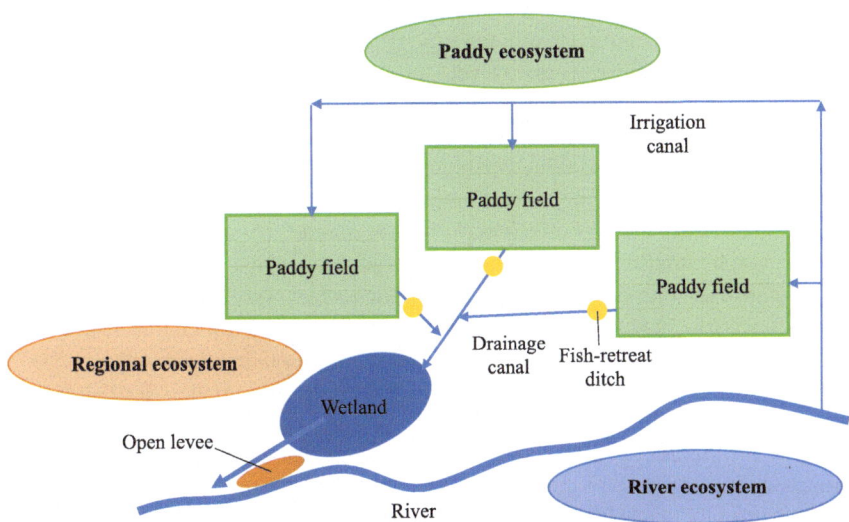

Fig. 15.2 Desirable regional ecosystem

monsoon region. People all over the world must understand the importance of these regional ecosystems.

15.3   Importance of Basin Resident Dependence on Rivers

It is important to be aware of disaster prevention and maintenance. Wetlands are in the process of transitioning and will become dry land if not appropriately managed. Since managing wetlands is an effort with strong public interest, the government must provide financial support.

Above all, reestablishment of basin resident dependence on rivers is required. Due to the modernization of river technology, constructing embankments has become easier. If a solid embankment has been developed, then the area inland of the embankment is less likely to experience overflow during floods. As a result, it is no longer necessary for basin residents to cooperate during floods and take measures to minimize damage. It is presumed that the residents of a basin are less involved with aspects of the river, and the dependence of residents on that river tends to disappear.

An open levee is a technology for protecting an area from flooding and is created on the assumption that floods will result in a river overflowing its banks; open levees are referred to as autonomous technology. However, hydraulic control by continuous dikes does not require residents to activity protect themselves; therefore, basin residents lose their ability to work together (Teramura and Okuma 2005).

It is possible that the damage caused by large-scale water disasters in recent years has expanded due to the loss of residents' dependence on river basins. By reducing the chances of getting close to a river, the threat of a river cannot be felt. Since residents are not interested in the rivers that flow around them, they do not notice the rise in the water levels. As a result, the evacuation of residents in a basin is delayed.

Conserving and improving open levees through the basin hydraulic control project currently advocated by the Ministry of Land, Infrastructure, Transport and Tourism of Japan seems to be effective in restoring the residents' dependence on river basins. The conventional way of thinking will change hydraulic control from trapping a flood to addressing the overflowing water. For this basin hydraulic control to be successful, the dependence of residents on rivers must be restored.

However, hydraulic flood control measures such as open levees that are based on topographical conditions can result in inequalities among residents depending on the region they are located. To resolve this problem, it is essential for engineers and local residents to discuss these issues with compassion for the other person (Okuma 2004). Farmers must also manage the water levels at their drainage outlets for fish-retreat ditches.

Based on the above information, it is important for basin residents (including farmers and nonfarmers) to cooperate with each other to protect lives from large floods. Currently, when individuals are respected, there is resistance to the demand for socialization and standardization. Proper use of both of these as needed is required for future social development.

15.4 Wetland Management

It is important for local residents to meet to discuss wetland management. Participants should include basin residents, municipal government members and those with academic experience. With continued discussions, consensus will be reached (Fig. 15.3).

Box 15.1 Oriental White Stork, Ciconia Boyciana, Nests and has Three Chicks in Obama City

The Oriental white stork *Ciconia boyciana* nested for the first time in 55 years in Kunitomi District, Obama City, Fukui Prefecture, Japan. Nesting was last confirmed in Kunitomi District in May 1966. The Kunitomi District is the last place in Japan where the nesting of wild *Ciconia boyciana* has been confirmed, and three chicks hatched in 2021 (Appendix Fig. 15.4). The fish-retreat ditch introduced in this book was created in this area, and it is possible that its effects are being demonstrated through this nesting. By linking wetlands, fish-retreat ditches and rivers, an environment where *Ciconia boyciana* could inhabit and nest was created. In addition, an open levee was installed in this area. If a wetland is created here, then it would be effective in reducing damage during floods. Thus, the Kunitomi District is considered to be a resilient area with abundant biodiversity and excellent flood control measures.

Fig. 15.3 Information flow among citizens, government and those with academic experience. *Source* Modified from Matsui (2015). Copyright 2015 Ecology and Civil Engineering Society

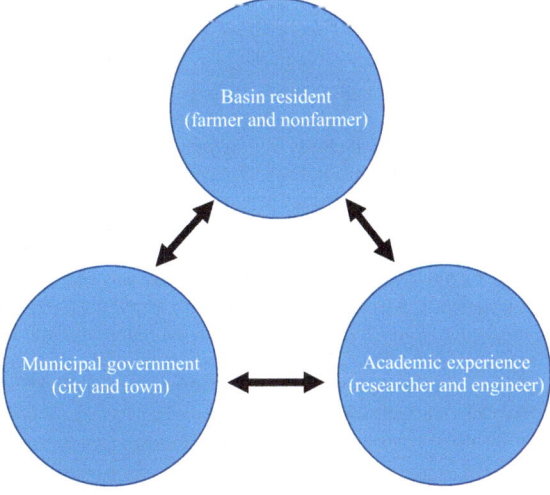

Fig. 15.4 Parent *Ciconia boyciana* and three chicks nesting in an artificial nest tower in Obama City, Fukui Prefecture, Japan. *Source* Township Promotion Association of Oriental White Stork, Kunitomi District, Obama City, Fukui Prefecture, Japan

Appendix

See Fig. 15.4.

References

Matsui A (2015) What it takes to practice better social capital improvement (in Japanese with English Abstract). Ecol Civ Eng 17:105–108. https://doi.org/10.3825/ece.17.105

Moriyama H, Iijima H, Harada N (1990) A favorable distribution pattern of aquatic biotopes in terms of the dispersal ability of dragonflies (in Japanese). Man Environ 15:2–15

Nishida K, Fujii C, Minagawa A, Senga Y (2006) Research on migration and dispersal range of freshwater fish that reproduce in temporary water area -Case study of Mukojima-channel in Hino-city and Fuchu-channel in Kunitachi-city, Tokyo- (in Japanese with English Abstract). Trans Japan Soc Irrig, Drainage Reclam Eng 244:151–163. https://doi.org/10.11408/jsidre1965.2006.553

Okuma T (2004) Creating local ideas 1 Technology also has autonomy -Tradition and modernity of water control technology-(in Japanese). Rural Culture Association Japan, Tokyo

Teramura J, Okuma T (2005) A study on evolution and role of open levees on alluvial-fan rivers in the Hokuriku District -from the view point of decentralization of river-engineering decision making- (in Japanese with English Abstract). J Hist Stud Civil Eng 24:161–171. https://doi.org/10.11532/journalhs2004.24.161

The manufacturer's authorised representative in the EU is Springer
Nature Customer Service Centre GmbH, Europaplatz 3, 69115 Heidelberg,
Germany. If you have any concerns regarding our products, please
contact ProductSafety@springernature.com

Printed and bound by CPI Group (UK) Ltd, Croydon, CR0 4YY

29/04/2026

02099528-0002